Spreadsheet

Nuts & Bolts

Pi Luna

ISBN: 978-0-9986860-0-4

Note about Technology

There are many different spreadsheet programs: Microsoft Excel®, Google Sheets®, Numbers for Mac®, Apache Open Office®, etc. While every spreadsheet application has unique features, *Spreadsheet Nuts & Bolts* covers universal features shared by the most common spreadsheet applications. Since technology is constantly evolving, some of the instructions in this book may become out of date. Visit your application's website for the most up-to-date information.

Table of Contents

Chapter 1: Getting Started

Whether you use spreadsheets at work, in school, or for personal reasons, spreadsheet applications serve a variety of purposes. In this chapter, learn how to navigate through the parts of a spreadsheet and stay organized.

Chapter 1 Contents

Topics	Examples
1.1 Parts of a Spreadsheet	Example 1.1A Using a Mouse to Select Cells Example 1.1B De-Selecting a Range of Cells Example 1.1C Using a Touchscreen to Select Cells
1.2 Navigation	Example 1.2A Using the Arrow Keys to Scroll Example 1.2B Scrolling Up and Down Example 1.2C Scrolling Side-to-Side Example 1.2D Zooming In and Out
1.3 Shortcuts	Example 1.3A Determining the Function of an Icon
1.4 Organization	Example 1.4A Renaming Worksheets Example 1.4B Adding New Worksheets

1.1 Parts of a Spreadsheet

A spreadsheet is a table comprised of columns, rows, and cells, which are used to organize data, make calculations, and store information.

Identifying Parts of a Spreadsheet

Columns

Columns are labeled with the letters of the alphabet and run vertically.

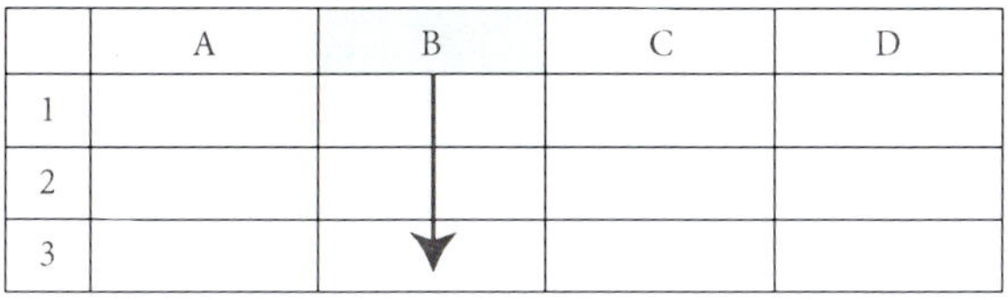

Rows

Rows are labeled with numbers and run horizontally.

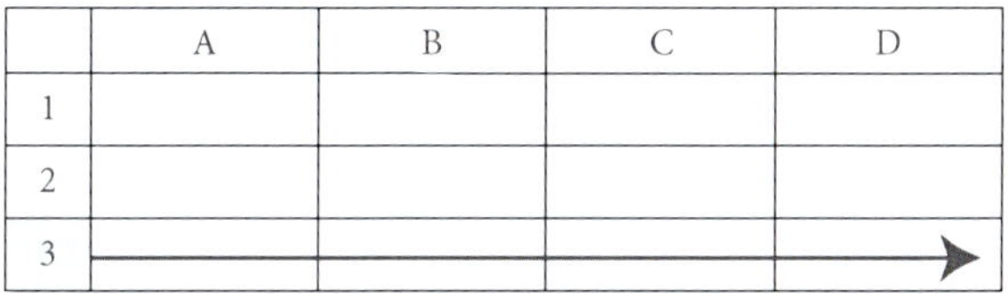

Cells

A cell is the intersection of a row and a column.

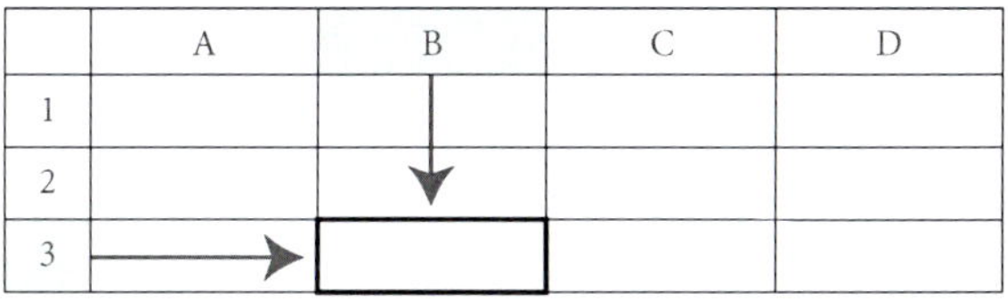

Remembering Which is Which

It is easy to mix up columns and rows. Think about rows as rowing a boat, because a boat moves from side-to-side. Columns are used in architecture to support buildings, and are built vertically.

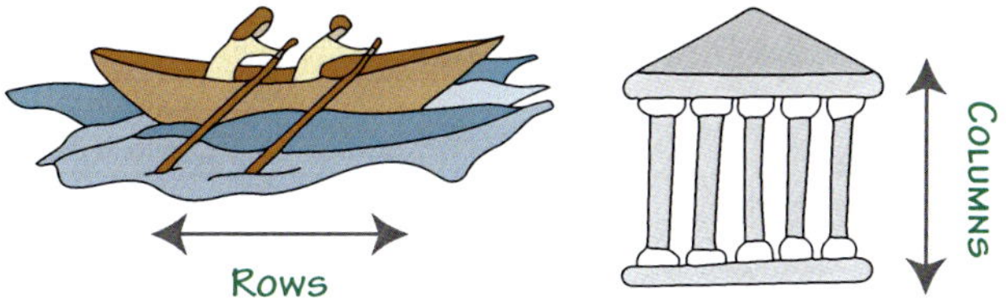

Cell Selection

When a cell or range of cells is selected, it will be highlighted. Once selected, you can enter text, numbers, or formulas into a cell.

Selection Tools

You can select cells using a mouse or a touchscreen.

Cursor

The cursor is used to make selections on the screen.

Mouse

The mouse is the physical device used to move the cursor on the screen.

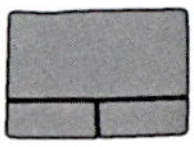

Trackpad

The trackpad is a touch device used to move the cursor on the screen.

Touchscreen

The touchscreen is an interface that uses your finger in place of the cursor.

Using a Mouse

Use a mouse to select cells. *Note: The features on a trackpad are similar to those on a mouse.*

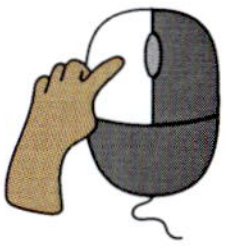

Click

Click to select a cell.

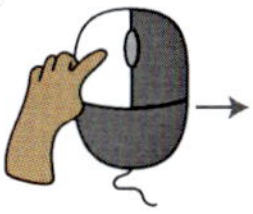

Click, Hold, and Drag

Click, hold, and drag the mouse to select multiple cells.

Double-Click

Double-click to edit the contents of a cell.

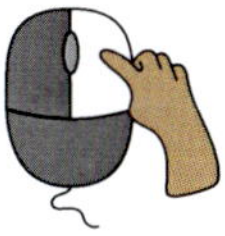

Right-Click

Right-click for formatting and editing options.

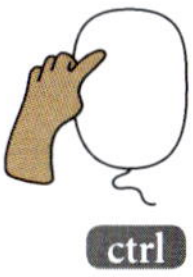

Right-Click on a Mac

If your Mac mouse does not have a right-click button, press and hold the ctrl key while clicking the mouse.

Example 1.1A Using a Mouse to Select Cells

Single Cell

Click on a cell to select it.

	A	B	C	D
1				
2				
3				

Column

Click on a column letter to select an entire column.

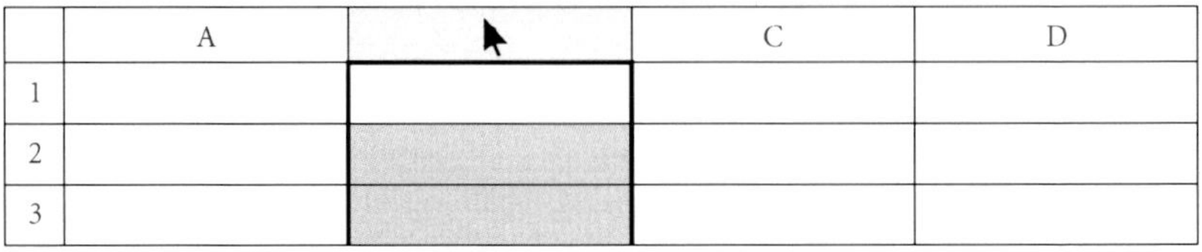

Row

Click on a row number to select an entire row.

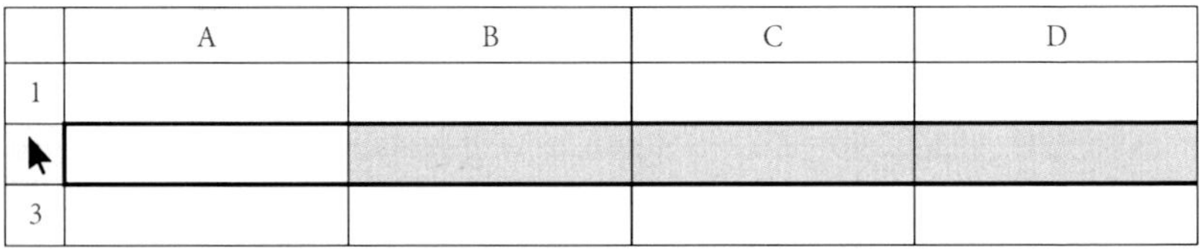

Range of Cells

Click and hold on the first cell and drag to the last cell in the selection. Then release the mouse. *Note: In some spreadsheet applications, the first cell will not become highlighted even though it is selected.*

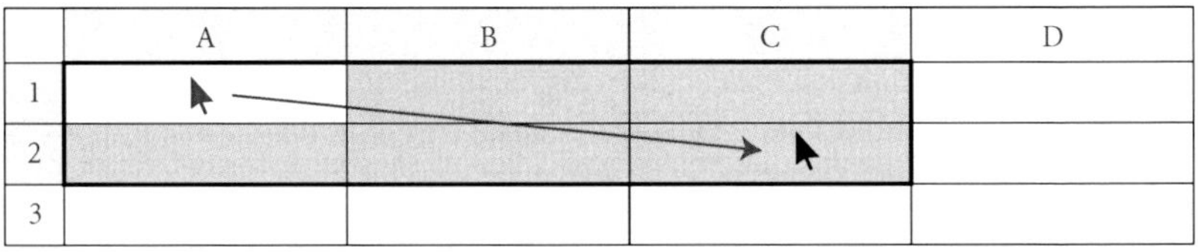

Non-Adjacent Cells

Press and hold the ctrl key on the keyboard while clicking on the desired cells. *Note: On a Mac, press and hold the* command *key instead of the* ctrl *key.*

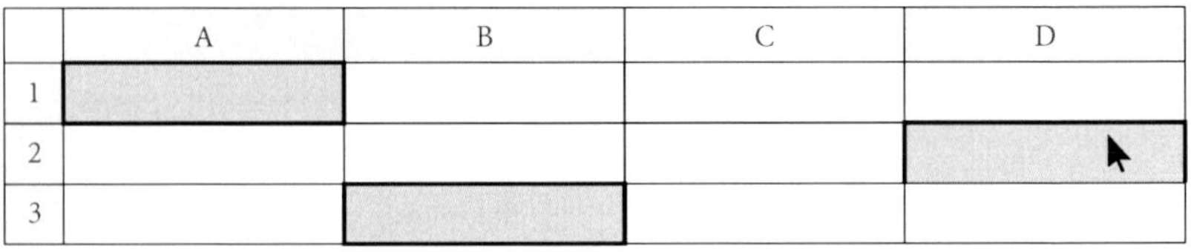

Getting Out of a Selection

If a group of cells is selected, use the arrow keys or your cursor to get out of the selection.

Example 1.1B De-Selecting a Range of Cells

Use the Arrow Keys

Press one of the arrow keys to get out of a selection.

Click on a Blank Cell

Click on a blank cell to get out of the selection.

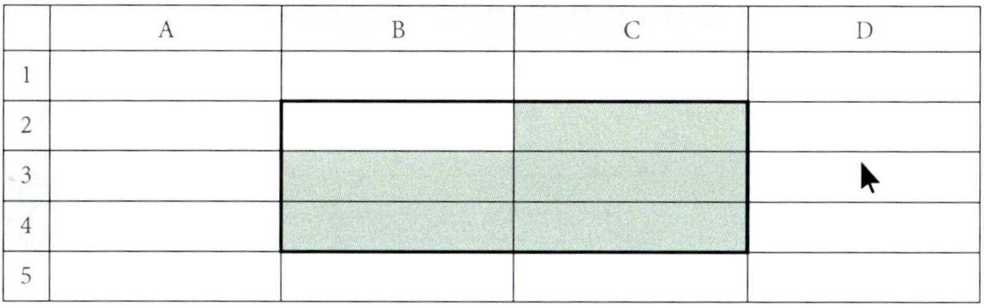

Do Not Use the Enter or Tab Keys

The enter or tab key on your keyboard will not get you out of a group of selected cells.

Using Touchscreen Devices

Spreadsheet applications work on many devices, including touchscreens. Most touchscreen features work the same as they do with a mouse, but there are some differences.

Tap

Tap the screen and lift your finger to select a cell.

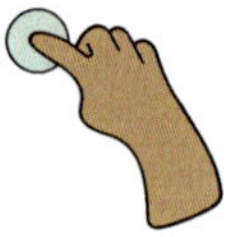

Press and Hold

Press and hold your finger on the screen to open menu options. This is equivalent to right-clicking on a mouse.

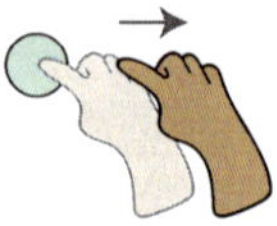

Press, Hold, and Drag

Press, hold, and drag your finger across the screen to select a range of cells.

Example 1.1C Using a Touchscreen to Select Cells

The process of selecting a range of cells on a touchscreen is slightly different from using a mouse.

Step 1

Open a new spreadsheet and tap inside the first cell to select it.

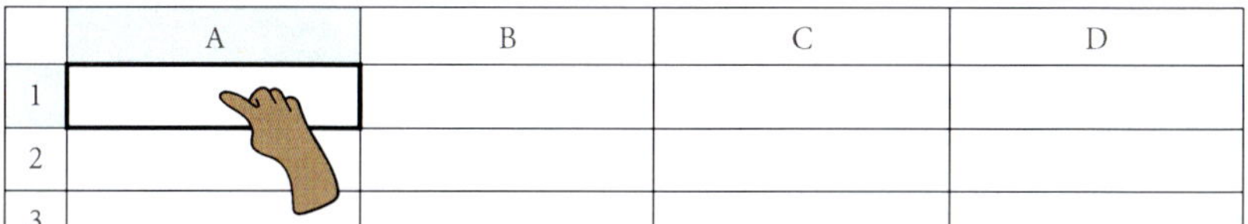

Step 2

Circles will appear on two corners of the cell.

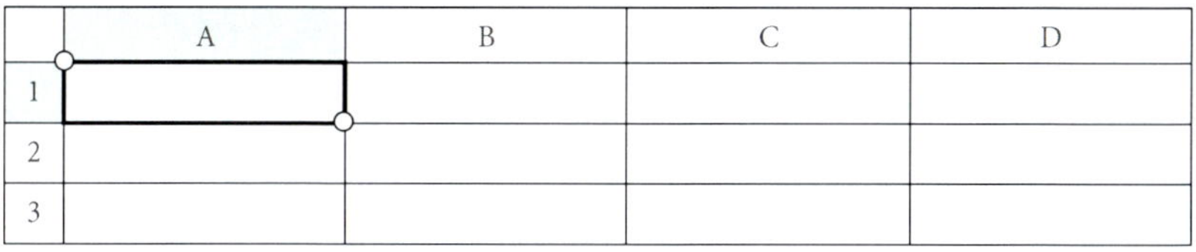

Step 3

Press one of the corners, hold and drag to reach your desired selection.

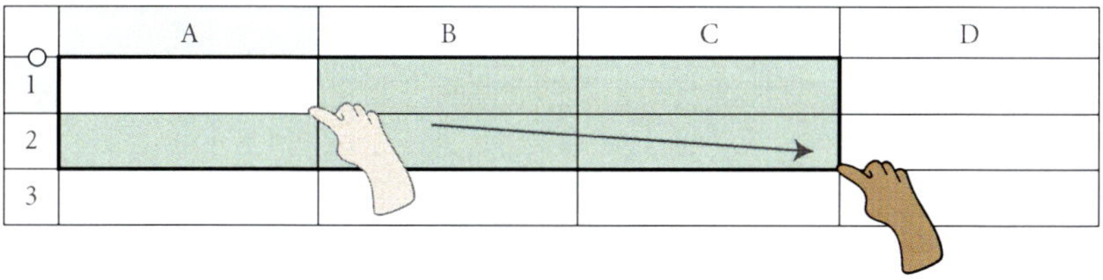

Non-Adjacent Cells on a Touchscreen

Typically, touchscreens cannot be used to select non-adjacent cells.

1.2 Navigation

Spreadsheets are versatile in their layout and organization, and learning to navigate them efficiently will improve your experience. Moving around a spreadsheet is often referred to as *scrolling*.

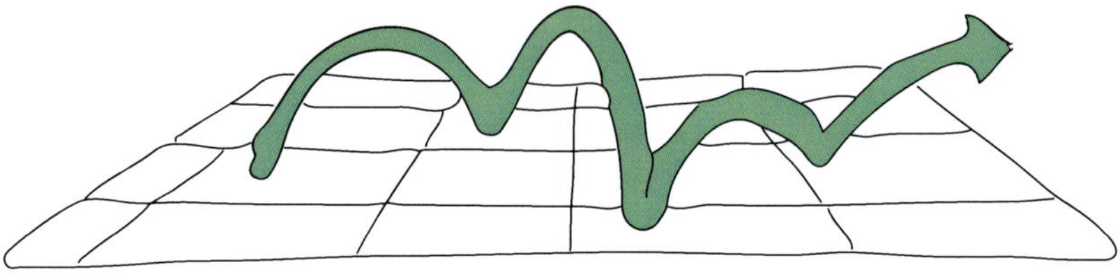

Example 1.2A Using the Arrow Keys to Scroll

In addition to using a mouse or touchscreen features, you can move from one cell to the next by pressing the arrow keys on your keyboard.

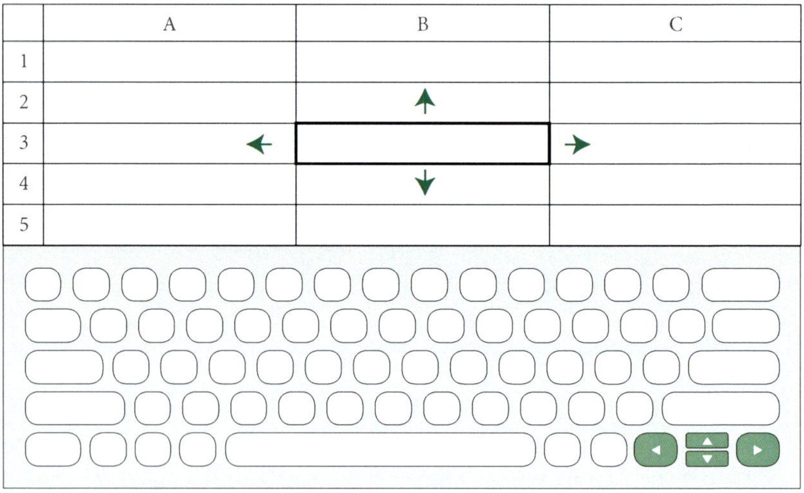

Example 1.2B Scrolling Up and Down

The enter key can be used to scroll up and down in a spreadsheet. *Note: The* return *key on a Mac is the same as the* enter *key on a PC.*

Scroll Down

Press the enter key to scroll down one cell.

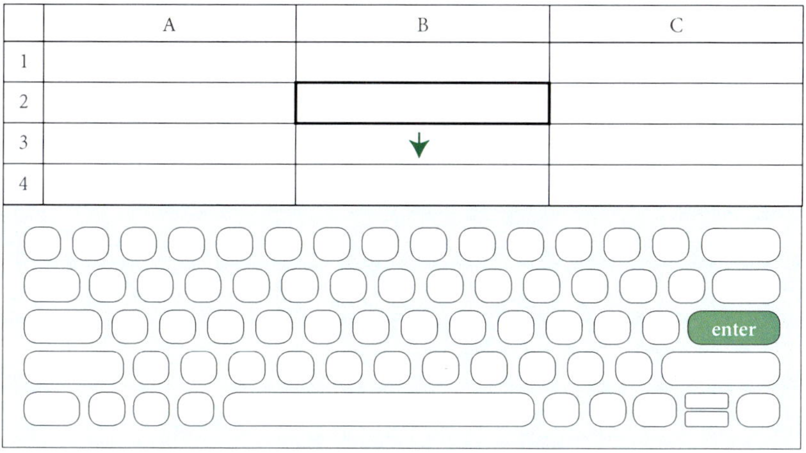

Scroll Up

Hold shift and press enter to scroll up one cell.

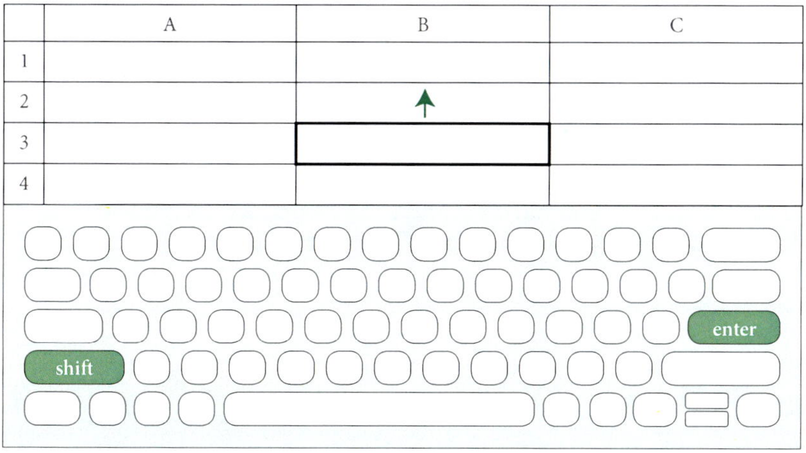

Example 1.2C Scrolling Side-to-Side

The [tab] key scrolls side-to-side in a spreadsheet.

Scroll Right

Press [tab] to scroll right one cell.

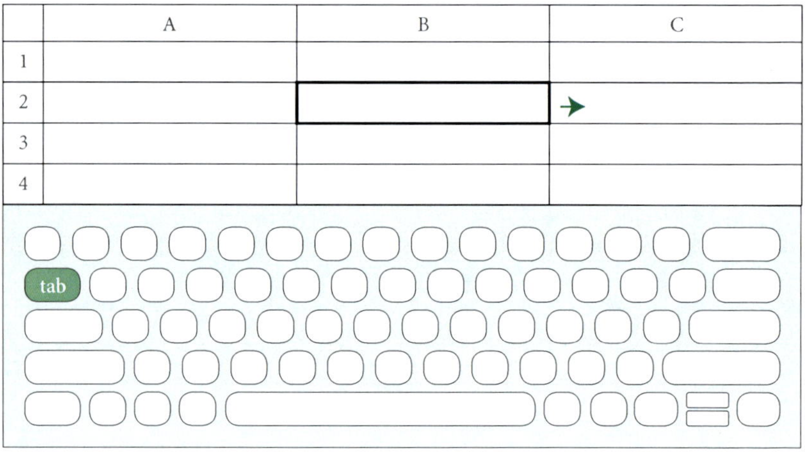

Scroll Left

Hold [shift] and press [tab] to scroll left one cell.

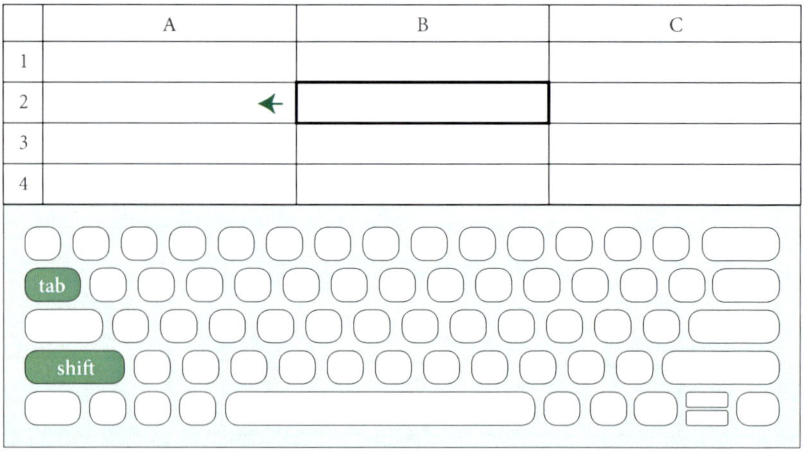

Example 1.2D Zooming In and Out

Being able to adjust the view of your spreadsheet is helpful. In this example, adjust the view using Microsoft Excel® or Google Sheets®.

Using Microsoft Excel®

Use the slider (near the bottom of the screen) to zoom in and out.

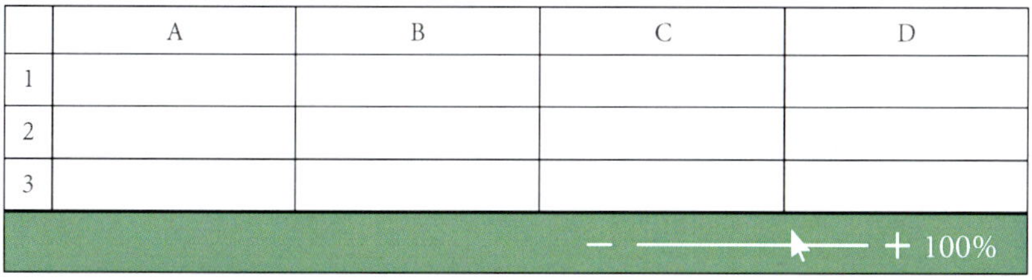

Using Google Sheets®

Hold [ctrl] and press the plus [+=] or minus [–-] keys on your keyboard to zoom in and out. *Note: On a Mac, hold the* [command] *key instead of the* [ctrl] *key.*

1.3 Shortcuts

Spreadsheet applications come with many icons, which are useful shortcuts for formatting. The location and appearance of icons changes, so hover over an unfamiliar icon to see what it does.

Example 1.3A Determining the Function of an Icon

In this example, hover the cursor over an icon to see its function.

Step 1

Open a new spreadsheet and hover your cursor over an icon.

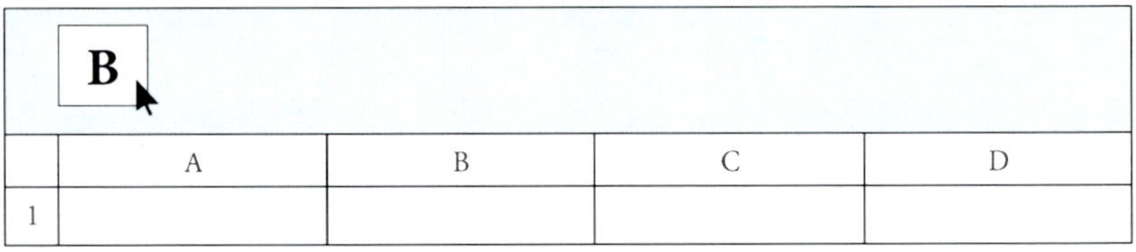

Step 2

A label will appear next to the icon telling you its function.

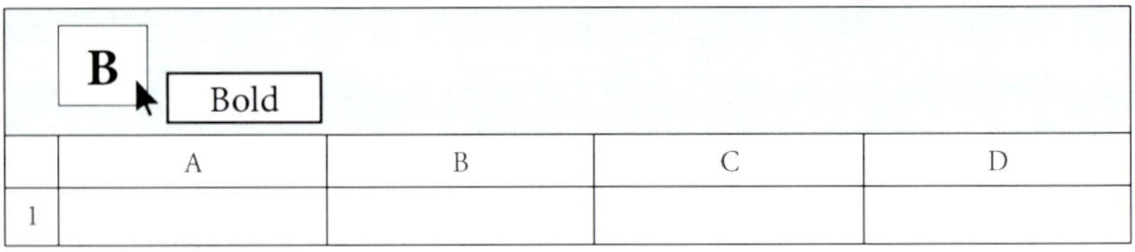

Keyboard Shortcuts

Many of the most commonly used icons and menu options have keyboard shortcuts that can be accessed by pressing a combinations of keys. Including special characters, there are hundreds of keyboard shortcuts. This book covers the most common shortcuts to cut down on time spent navigating with a mouse.

Modifier Keys

The shift and ctrl keys on a keyboard are used for a variety of keyboard shortcuts in a spreadsheet application.

Shift Key

The (shift) key is the most common shortcut. It is used to either produce a capital letter or access the secondary function of a key such as the percentage (%) symbol.

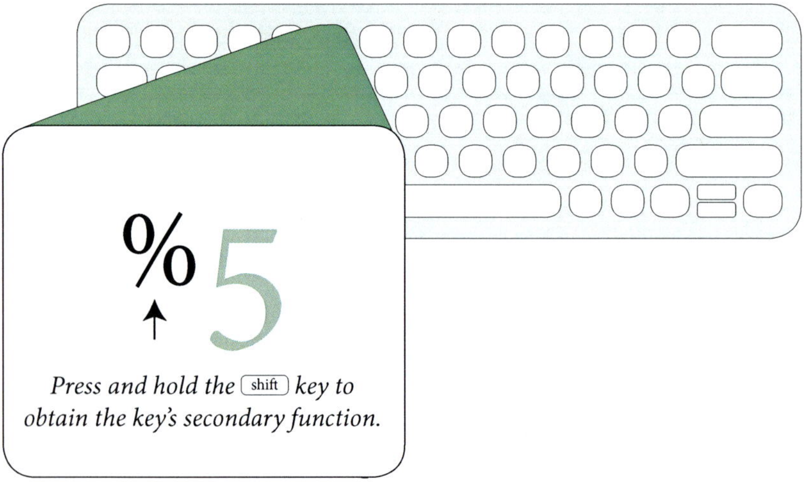

Control Key

The (ctrl) key provides quick access to features, such as copy, paste, and save. The following common shortcuts work in spreadsheets and many other applications.

Editing

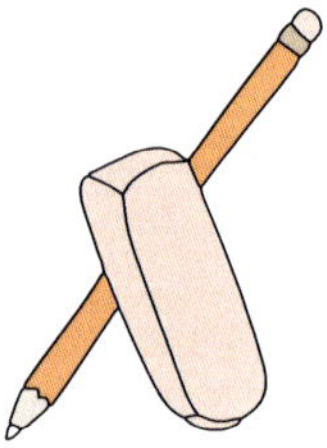

Cut	ctrl x
Copy	ctrl c
Paste	ctrl v
Undo	ctrl z
Redo	ctrl shift z
Select All	ctrl a

Files

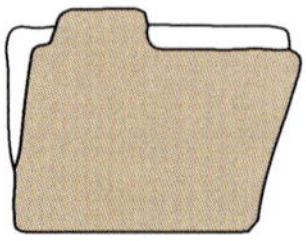

Save	ctrl s
Open	ctrl o
Print	ctrl p

Modifier Keys on a Mac

The modifier keys on a Mac keyboard are slightly different.

Use the ctrl *key to right-click on a Mac.*

Use the command *key on a Mac instead of the* ctrl *key on a PC for keyboard shortcuts.*

1.4 Organization

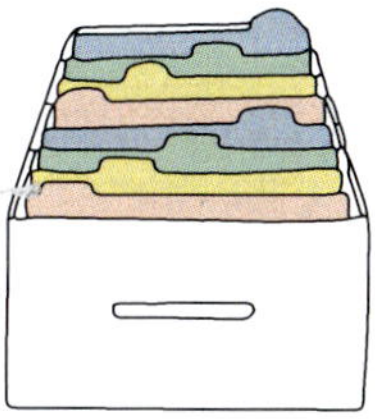

Staying organized is useful when you need to review a topic you have already worked on. Spreadsheet applications have features to help you stay organized.

Worksheets in a Workbook

A great way to keep your spreadsheet files organized is to understand the difference between worksheets and workbooks.

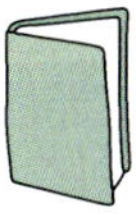

Workbook = File

A workbook is a spreadsheet file.

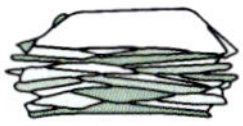

Worksheet = Page

Worksheets are the spreadsheets contained within a workbook. Use one or multiple worksheets, depending on your needs.

Example 1.4A Renaming Worksheets

In this example, change the default name "Sheet 1" to "New Name."

Step 1

Open a new spreadsheet and double-click on the default worksheet name "Sheet 1" located at the bottom of the spreadsheet.

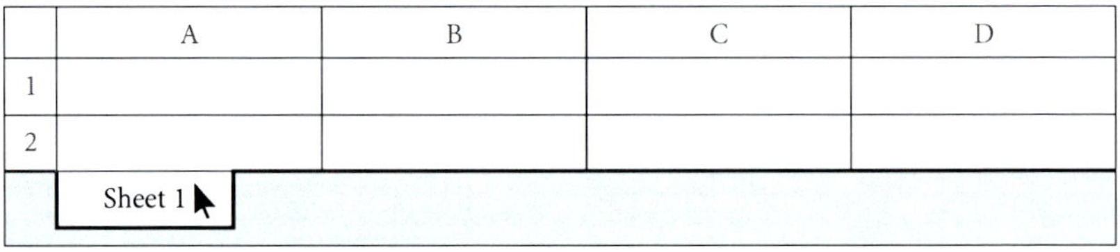

Step 2

It will become highlighted.

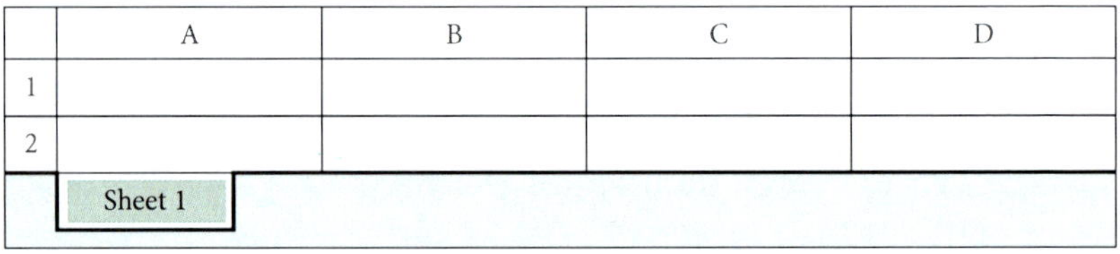

Step 3

Type the new sheet name.

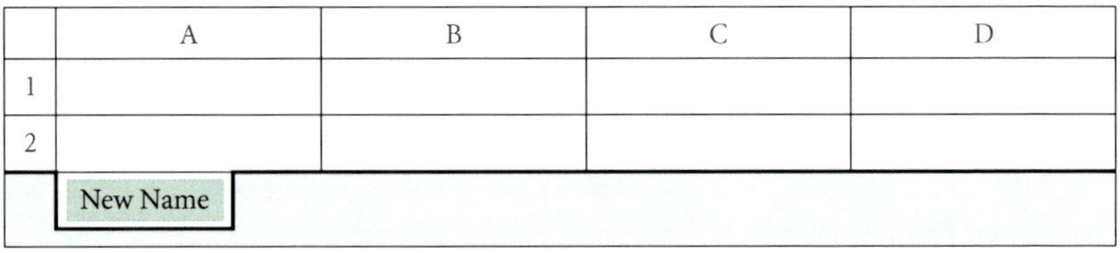

Step 4

Press enter to complete the name change.

Result

The worksheet will be renamed.

	A	B	C	D
1				
2				
	New Name			

Example 1.4B Adding New Worksheets

Consider adding new worksheets within a workbook to stay organized. This is particularly helpful when comparing several related options or scenarios.

Step 1

Open a new spreadsheet and click on the plus (+) symbol next to the sheet name.

	A	B	C	D
1				
2				
	Sheet 1	+		

Result

A new worksheet will be added with a default name, such as “Sheet 2.” Add as many worksheets as needed and rename them to stay organized.

	A	B	C	D
1				
2				
	Sheet 1	Sheet 2		

Effective Sheet Names

Effective names make workbooks and worksheets more useful for future reference. Generic or common names can make it difficult to locate the sheet you are looking for. Consider using names that are clear, specific, and distinct.

Examples of Effective Sheet Names

	A	B	C	D
1				
2				
	Car Loan_5 years	$18,000 at 3.5% 48 months		Math 4B #22 p. 82

Clear

Specific

Distinct

Examples of Ineffective Sheet Names

	A	B	C	D
1				
2				
	Last problem in class	Options		#17

Unclear

Not Specific

Not Distinct

Chapter 1 Summary

Spreadsheets applications can be used for business, academic, and personal purposes. This chapter covered the basics of how to get started, navigate the parts of a spreadsheet, use important shortcuts, and stay organized.

Chapter 2: Text

Spreadsheets allow you to work with text and there are several important features that will make your work visually appealing and easier to edit. In this chapter, learn how to format and edit text inside a spreadsheet.

Chapter 2 Contents	
Topics	Examples
2.1 Format Text	Example 2.1A Formatting Text
2.2 Edit Text	Example 2.2A Overwriting Text Example 2.2B Editing Text Example 2.2C Deleting Text
2.3 Fit Text	Example 2.3A Merging Cells Example 2.3B Wrapping Text Example 2.3C Adjusting Column Width
2.4 Manage Rows & Columns	Example 2.4A Inserting Rows & Columns Example 2.4B Deleting Rows & Columns

2.1 Format Text

Spreadsheets can look complex, so consider formatting headings and key pieces of information to help guide the eye and make working with the information easier. The following are just a few of the many formatting options available, depending on your needs and the specific spreadsheet application.

Emphasizing Text

Bold, italics, and underlining are some of the main ways to emphasize particular cells or text.

Bold   Italics   Underline

Aligning Text

Text will be aligned to the left of the cell by default, but this can be changed to suit your needs.

Left   Center   Right

Colors and Borders

The text color, the cell color and the borders can be changed.

Cell Color   Text Color   Cell Borders

Example 2.1A Formatting Text

In this example, format the word "Text" to be bold and center aligned.

Step 1

Open a new spreadsheet and type the word "Text" in the first cell.

	A	B	C	D
1	Text			
2				

Step 2

Press enter, tab, or click on a blank cell to complete the entry.

Step 3

Click on the cell with word “Text.”

	A	B	C	D
1	Text			
2				

Step 4

While the cell is still selected, click on the bold icon at the top of the window.

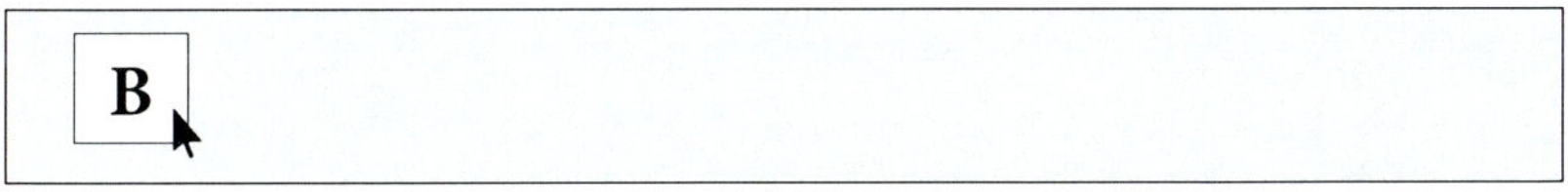

Step 5

Click on the center align icon.

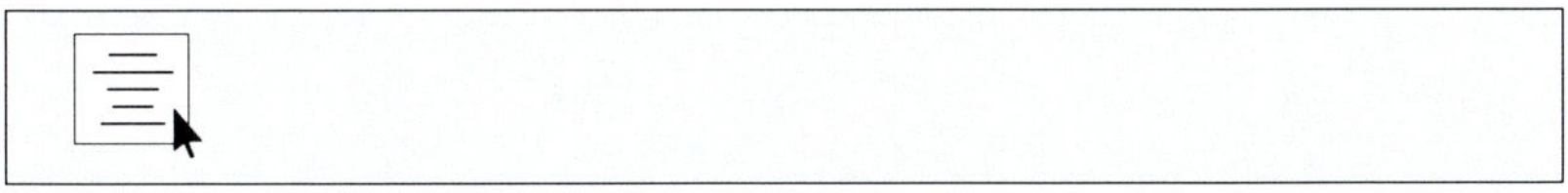

Result

The word “Text” will become bold and center aligned.

	A	B	C	D
1	**Text**			
2				

2.2 Edit Text

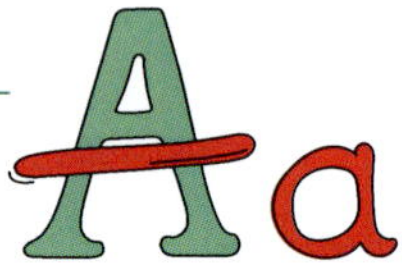

There are several ways to edit text inside a spreadsheet. You can edit existing text or completely overwrite text.

Ways to Edit Text

Overwriting Text

Replaces cell contents with new text.

Editing Text Inside a Cell

Uses a cursor to edit existing text.

Deleting Text

Deletes the text inside a cell or range of cells.

Example 2.2A Overwriting Text

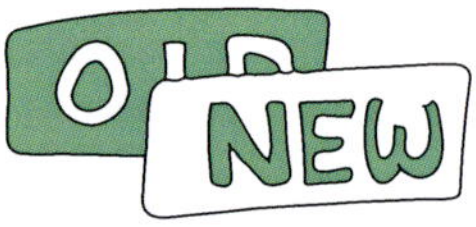

One way to edit text is to overwrite it. In this example, change the word "Original" to "New" inside a selected cell.

Step 1

Open a new spreadsheet and type the word "Original" in the first cell.

	A	B	C	D
1	Original			
2				

Step 2

Press enter, tab, or click on a blank cell to complete the entry.

Step 3

Click on the cell that contains the text "Original."

	A	B	C	D
1	Original			
2				

Step 4

Type the new text.

	A	B	C	D
1	New			
2				

Step 5

Press enter, tab, or click on a blank cell to complete the change.

Result

The word "Original" is replaced with the word "New."

	A	B	C	D
1	New			
2				

Example 2.2B Editing Text

You can also edit the text already in a cell. In this example, change the word, "New" to the phrase, "New Text."

Step 1

Open a new spreadsheet and type the word "New" in the first cell, or continue the previous example.

	A	B	C	D
1	New			
2				

Step 2

Press enter, tab, or click on a blank cell to complete the entry.

Step 3

Double-click on the cell you want to edit.

	A	B	C	D
1	New			
2				

Step 4

A blinking text cursor will show up in the text. Use the arrow ◄ ► keys or the mouse to move the cursor to the desired location to edit the text.

	A	B	C	D
1	NeIw			
2				

Step 5

Once the cursor is in the desired location, type the new text.

	A	B	C	D
1	New Text			
2				

Step 6

Press enter, tab, or click on a blank cell to complete the change.

Result

The word "New" is changed to the phrase, "New Text."

	A	B	C	D
1	New Text			
2				

Example 2.2C Deleting Text

As you develop spreadsheets, it may be necessary to delete the text inside cells. You can delete single cells, a selection of multiple cells, or entire rows or columns. In this example, delete the word "Text" in a selection of cells.

Step 1

Open a new spreadsheet and type the word "Text" in the first three cells of row 1.

	A	B	C	D
1	Text	Text	Text	
2				

Step 2

Select the cells with the contents you want to delete.

	A	B	C	D
1	Text	Text	Text	
2				

Step 3

Press the delete key. If you are using Google Sheets®, backspace will also work.

Result

The content in the cells will be deleted.

	A	B	C	D
1				
2				

2.3 Fit Text

Default cell sizes do not accommodate a lot of text. When this happens you have three options: merge several cells together, use text wrapping, or adjust the column width.

Ways to Fit Text

There are three main ways to create room for oversized text.

Merge Cells
Combines multiple cells into one larger cell.

Wrap Text
Increases the row height so the text fits on multiple lines.

Increase Column Width
Increases the column width so the text fits in a single cell.

Example 2.3A Merging Cells

The first option to fit oversized text is to merge cells. This option combines multiple cells into one larger cell. In this example, merge two cells to create room for the phrase, "The text won't fit in this cell."

Step 1

Open a new spreadsheet and type the phrase, "The text won't fit in this cell."

	A	B	C	D
1	The text won't fit	in this cell.		
2				

Step 2

Press enter, tab, or click on a blank cell to complete the entry.

Step 3

Select the cells that contain text.

	A	B	C	D
1	The text won't fit in this cell.			
2				

Step 4

Click on the merge icon located near the top of your window.

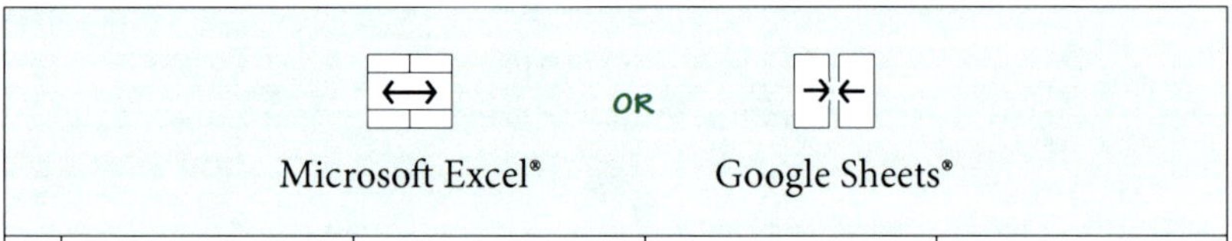

	A	B	C	D
1				

Result

As a result, the two cells will be merged into one cell.

	A	B	C	D
1	The text won't fit in this cell.			
2				

Example 2.3B Wrapping Text

The second option to fit oversized text is text wrapping. This option maintains the column width and increases the row height to fit text on several lines. In this example, wrap text to create room for the phrase, "The text won't fit in this cell."

Step 1

Open a new spreadsheet and type the phrase, "The text won't fit in this cell."

	A	B	C	D
1	The text won't fit in this cell.			
2				

Step 2

Press enter, tab, or click on a blank cell to complete the entry.

Step 3

Click on the cell where the text begins.

	A	B	C	D
1	The text won't fit	in this cell.		
2				

Step 4

Click on the text wrap icon located near the top of your spreadsheet application.

Microsoft Excel® Google Sheets®

	A	B	C	D
1				

Result

As a result, the row height will increase to fit the text on more than one line.

	A	B	C	D
1	The text won't fit in this cell.			
2				

Example 2.3C Adjusting Column Width

The third option to fit oversized text is to increase the column width. In this example, adjust the column width to accommodate the phrase, "The text won't fit in this cell" using a mouse or a touchscreen device.

Using a Mouse

Step 1

Open a new spreadsheet and type the phrase, "The text won't fit in this cell."

	A	B	C	D
1	The text won't fit	in this cell.		
2				

Step 2

Press enter, tab, or click on a blank cell to complete the entry.

Step 3

Place your cursor on the column border.

	A	B	C	D
1	The text won't fit	in this cell.		
2				

Step 4

Click and drag until you reach the desired width. *Note: You can also double-click the column border to auto-adjust the width.*

	A		B	C	
1	The text won't fit	in this cell.			
2					

Result

The column will adjust to the new width.

	A	B	C	
1	The text won't fit in this cell.			
2				

Using a Touchscreen

Step 1

Open a new spreadsheet and type the phrase, "The text won't fit in this cell."

	A	B	C	D
1	The text won't fit	in this cell.		
2				

Step 2

Press enter, tab, or click on a blank cell to complete the entry.

Step 3

Tap the column border to select it.

	A	B	C	D
1	The text won't fit	in s cell.		
2				

Step 4

Then press and hold the column border and drag until you reach the desired width.

	A		B	C	
1	The text won't fit	in this cell.			
2					

Result

The column will adjust to the new width.

	A	B	C	
1	The text won't fit in this cell.			
2				

2.4 Manage Rows & Columns

You can insert new rows and columns to add more text to your spreadsheet. You can also delete unwanted rows and columns.

Ways to Manage Rows & Columns

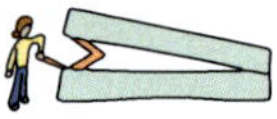

Insert Rows & Columns

Create room for more text.

Delete Rows & Columns

Remove unwanted text.

Example 2.4A Inserting Rows & Columns

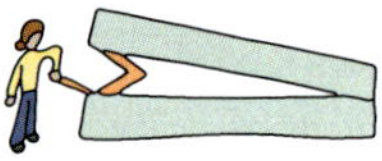

You can insert new rows and columns. In this example, insert a new row to add the name, "Alex" to a list of names.

Step 1

Open a new spreadsheet and create the following list of names.

	A	B	C	D
1	Adrian			
2	Alan			
3	Andrew			
4	Ann			

Step 2

In order to be in alphabetical order, the name "Alex" needs to go between "Alan" and "Andrew." Right-click on the row number of the desired location.

	A	B	C	D
1	Adrian			
2	Alan			
	Andrew			
4	Ann			

Step 3

A menu will appear. Click on insert.

	A	B	C	D
1	cut			
2	copy			
3	paste			
4	insert			
	delete			

Step 4

A new blank row will be inserted. Click on the cell you want to edit.

	A	B	C	D
1	Adrian			
2	Alan			
3				
4	Andrew			
5	Ann			

Step 5

Type the new name in the selected cell.

	A	B	C	D
1	Adrian			
2	Alan			
3	Alex			
4	Andrew			
5	Ann			

Step 6

Press enter, tab, or click on a blank cell to complete the change.

Result

The new name will be included in the list. *Note: Follow the same procedure for inserting a new column.*

	A	B	C	D
1	Adrian			
2	Alan			
3	Alex			
4	Andrew			
5	Ann			

Example 2.4B Deleting Rows & Columns

You can delete unwanted rows and columns. In this example, delete a row to remove the name, "Andrew" from a list of names.

Step 1

Complete the steps in "Example 2.4A Inserting Rows & Columns" on page 36. Right-click on the number of the row you want to delete.

	A	B	C	D
1	Adrian			
2	Alan			
3	Alex			
	Andrew			
5	Ann			

Step 2

A menu will appear. Click on delete.

	A	B	C	D
1	cut			
2	copy			
3	paste			
4	insert			
5	delete			

Result

The row will be deleted. *Note: Follow the same procedure for inserting a new column.*

	A	B	C	D
1	Adrian			
2	Alan			
3	Alex			
4	Ann			

Chapter 2 Summary

Spreadsheets are designed to accommodate many different kinds of information, including text. This chapter covered important features of formatting and editing text that will make your work visually appealing and easier to edit.

Chapter 3: Numbers

In addition to the editing techniques used with text, there are features specific to numbers, including formatting for decimals, percentages, and currency. In this chapter, learn to work with and format numbers in a spreadsheet.

Chapter 3 Contents

Topics	Examples
3.1 Format Numbers	Example 3.1A Formatting Decimals Example 3.1B Formatting as Percentages Example 3.1C Formatting as Currency
3.2 Clear Formatting	Example 3.2A Removing Cell Formatting
3.3 Create Lists	Example 3.3A Creating a List of Numbers

3.1 Format Numbers

Some common formatting options are available as icons to quickly format numbers. The most common number formats are decimal, percentage, and currency.

Ways to Format Numbers

Decimals

Format values as decimals to a desired place value.

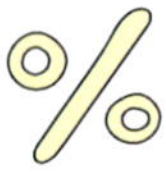

Percentages

Convert values to percentages.

Currency

Add dollar signs and round numbers to two decimal places.

Example 3.1A Formatting Decimals

Decimal numbers can repeat or continue with many place values. You can control how numbers are rounded in a spreadsheet. In this example, round a number to three decimal places.

Step 1

Open a new spreadsheet and type a decimal number to round.

	A	B	C	D
1	3.1875			
2				

Step 2

Press enter, tab, or click on a blank cell to complete the entry.

Step 3

Click on the cell to select it.

	A	B	C	D
1	3.1875			
2				

Step 3

While the cell is selected, click the appropriate icon to decrease the number of decimal places displayed.

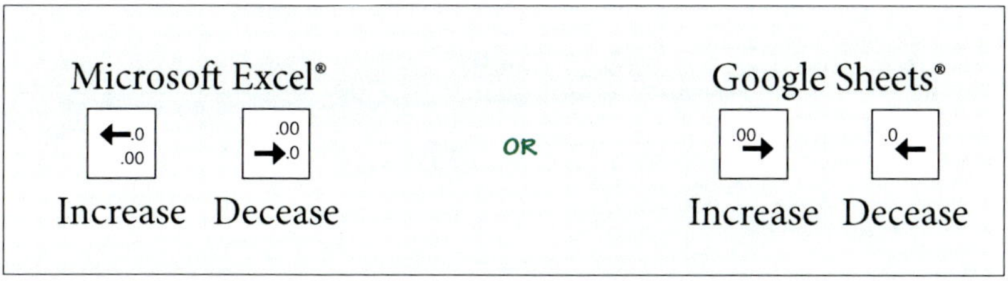

Result

The number will be rounded to three decimal places.

	A	B	C	D
1	3.188			
2				

Example 3.1B Formatting as Percentages

Spreadsheet applications can convert values to percentages. In this example, convert several values from a decimal to a percentage.

Decimal		Percentage
0.05	→	5%

Step 1

Open a new spreadsheet and type several numbers to convert to percentages.

	A	B	C	D
1	0.25	0.05	2	
2				

Step 2

Select the cells you want to convert to percentages.

	A	B	C	D
1	0.25	0.05	2	
2				

Step 3

Click the percentage % symbol near the top of the window.

Result

The numbers will be converted from decimal values to percentages. *Note: Use the place value icons to increase or decrease place values shown in the percentages.*

	A	B	C	D
1	25%	5%	200%	
2				

Example 3.1C Formatting as Currency

Using a spreadsheet is a great way to manage personal and business finances. In this example, change several numbers from decimal form to currency.

Step 1

Open a new spreadsheet and type several numbers to convert to currency.

	A	B	C	D
1	11.0877	22	0.5	
2				

Step 2

Select the cells you want to convert to currency.

	A	B	C	D
1	11.0877	22	0.5	
2				

Step 3

Click the dollar $ sign near the top of the spreadsheet application.

$

	A	B	C	D
1				

Result

The numbers will be rounded to 2 decimal places and formatted as currency.

	A	B	C	D
1	$11.09	$22.00	$0.50	
2				

Negative Numbers

When formatted as currency, negative numbers are often denoted with a negative sign. They can also be placed inside parentheses, in red text, or both.

	A	B	C	D
1	-$10.00	-$10.00		
2	($10.00)	($10.00)		

Fun Fact: Companies often say they are "in the black" to indicate a profit and "in the red" to indicate a loss. Black Friday is a day that retailers hope holiday-season spending will help them get back in the black.

3.2 Clear Formatting

Once a cell is formatted as currency or a percentage, it will keep the formatting even if the contents of that cell change. You can clear cell formatting.

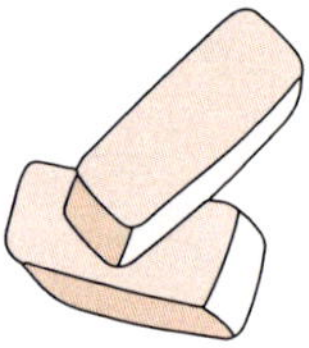

Example 3.2A Removing Cell Formatting

In this example, clear cell formatting to change it from a percentage back to its original decimal form using Microsoft Excel® or Google Sheets®.

Using Microsoft Excel®

Step 1

Complete the steps in "Example 3.1B Formatting as Percentages" on page 42. Select the cells where you want to clear the formatting.

	A	B	C	D
1	25%	5%	200%	
2				

Step 2

A menu near the top of the window displays the selection's formatting. Click on the menu.

percentage

	A	B	C	D
1	25%	5%	200%	

Step 3

Select general from the drop-down menu to clear the formatting.

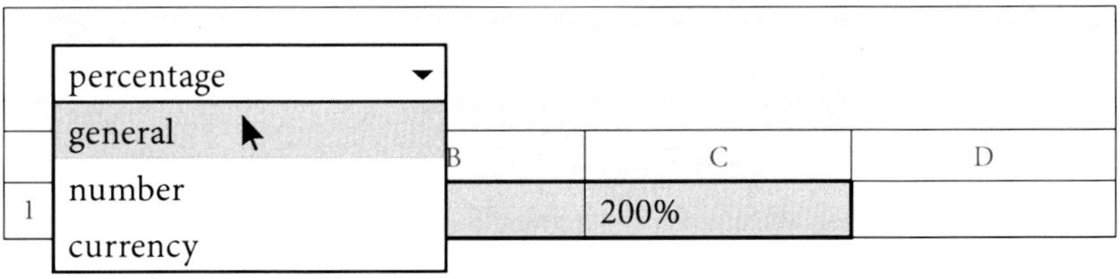

Result

The formatting will be cleared and the percentage will revert back to its decimal equivalent.

	A	B	C	D
1	0.25	0.05	2	
2				

Using Google Sheets®

Step 1

Complete the steps in "Example 3.1B Formatting as Percentages" on page 42. Select the cells where you want to clear the formatting.

	A	B	C	D
1	25%	5%	200%	
2				

Step 2

Click on the 123 icon near the top of the screen.

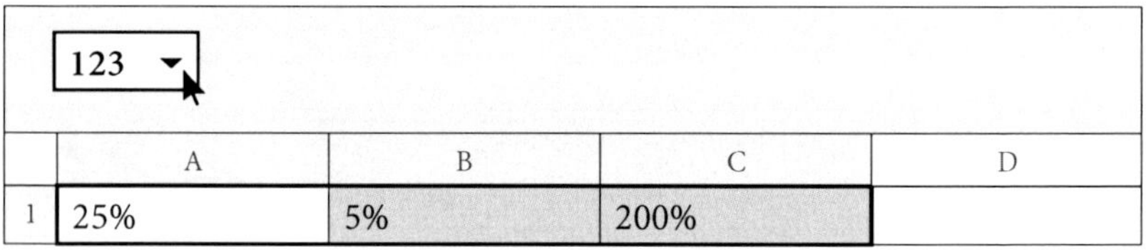

Step 3

Select automatic from the drop-down menu to clear the formatting.

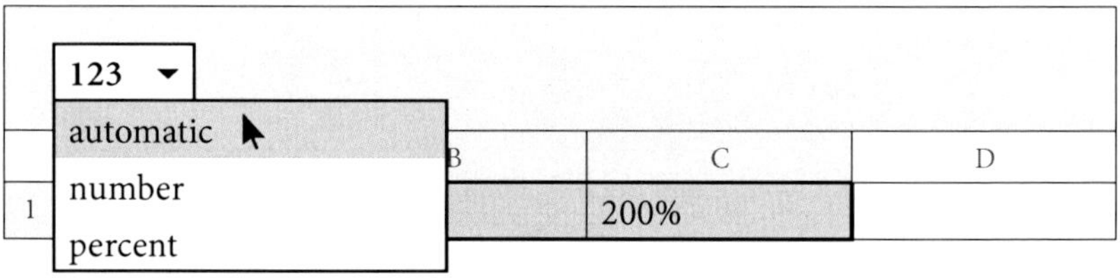

Result

The formatting will be cleared and the percentage will revert back to its decimal equivalent.

	A	B	C	D
1	0.25	0.05	2	
2				

3.3 Create Lists

When creating a list of numbers, your spreadsheet application can recognize and complete the numbering pattern for a list.

Example 3.3A Creating a List of Numbers

In this example, create a list of numbers from 1 to 5 using a mouse or a touchscreen device.

Using a Mouse

Step 1

Open a new spreadsheet and type the first two numbers in the list to establish a pattern.

	A	B	C	D
1	1			
2	2			
3				

Step 2

Select both cells.

	A	B	C	D
1	1			
2	2			
3				

Step 3

Click and hold the fill handle at the bottom right corner of the selection and drag down.

	A	B	C	D
1	1			
2	2			
3				

	A	B	C	D
1	1			
2	2			
3				
4				
5				

Result

Release the mouse and the list will fill in.

	A	B	C	D
1	1			
2	2			
3	3			
4	4			
5	5			

Using a Touchscreen

Step 1

Open a new spreadsheet and type the first two numbers in the list to establish a pattern.

	A	B	C	D
1	1			
2	2			
3				

Step 2

Select both cells. *Refer to "Example 1.1C Using a Touchscreen to Select Cells" on page 10 for instructions.*

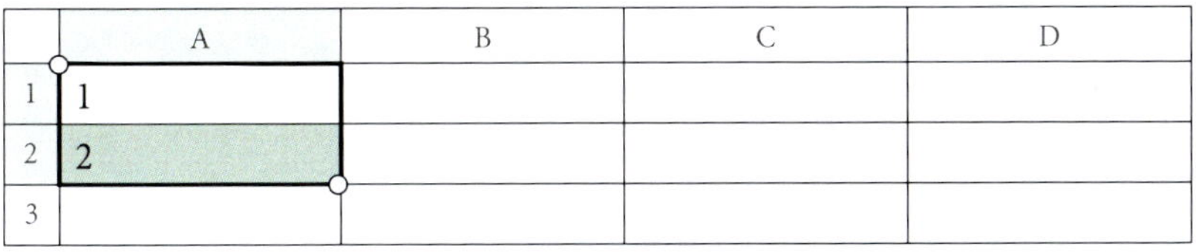

Step 3

Press and hold the bottom right corner of the selection.

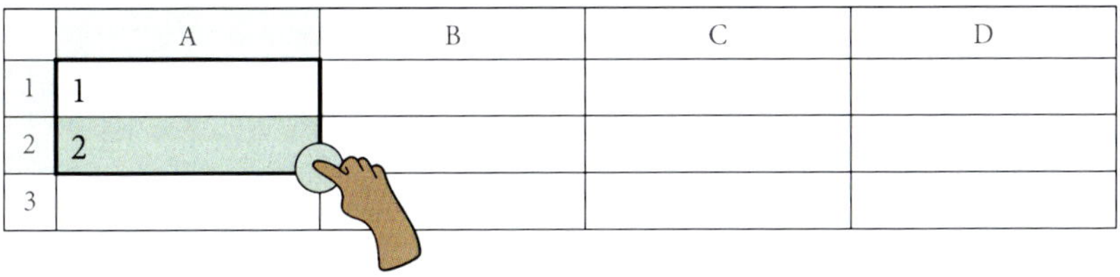

Step 4

A menu will appear. Choose the AutoFill option.

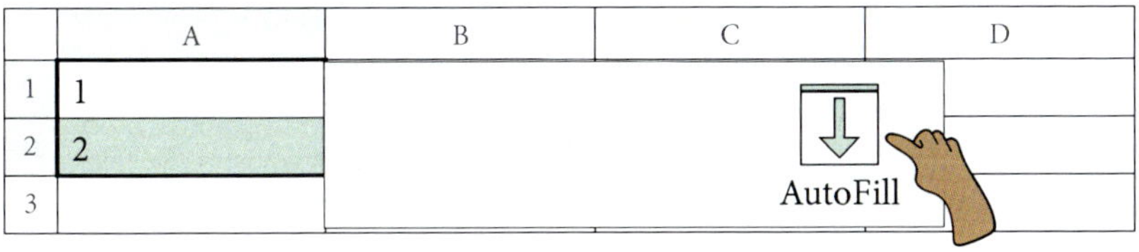

Step 5

An icon will appear on the bottom right corner of the cell. Press and hold the icon, and drag your finger down.

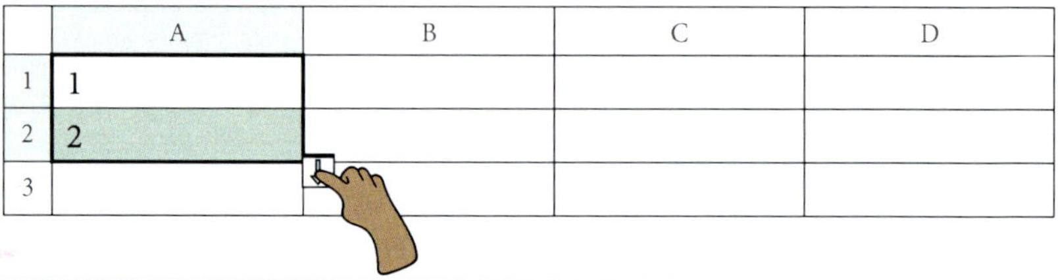

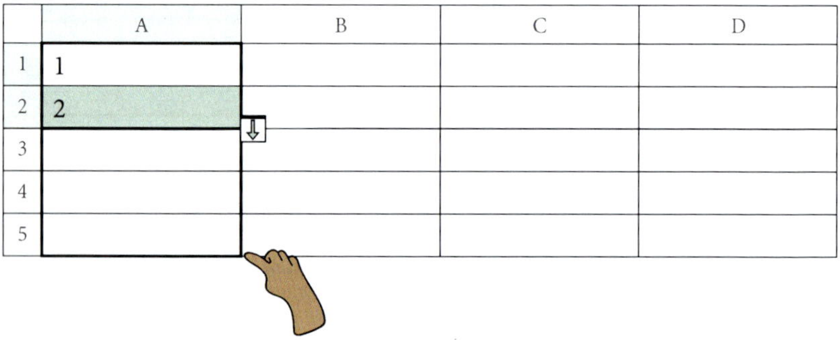

Result

Release your finger and the list will fill in.

	A	B	C	D
1	1			
2	2			
3	3			
4	4			
5	5			

Chapter 3 Summary

Many of the editing techniques used with text also apply to working with numbers; however, there are additional features specific to work with decimals, percentages, and currency. This chapter covered how to format numbers and create numbered lists in a spreadsheet.

Chapter 4: Formulas

Formulas are what make spreadsheets powerful. Like a calculator, a spreadsheet can be used to make calculations. You can make calculations in a single cell or use cell references to repeat calculations with multiple cells. In this chapter, learn to use formulas in spreadsheets.

Chapter 4 Contents

Topics	Examples
4.1 Build Formulas	Example 4.1A Building a Formula Example 4.1B Using the Formula Bar
4.2 Cell References	Example 4.2A Creating Cell References Example 4.2B Editing Referenced Cells Example 4.2C Using the Fill Handle Example 4.2D Creating a Fixed Cell Reference Example 4.2E Creating Sheet References
4.3 Copy & Paste Options	Example 4.3A Paste with Formulas Example 4.3B Paste Only Values

4.1 Build Formulas

Spreadsheets can quickly make many calculations using formulas, and every formula begins with an equals = sign.

Math Keyboard Shortcuts

When creating a formula, use the following keyboard shortcuts to obtain the appropriate math symbols.

Add	shift +==	Equals	+=
Subtract	_-	Greater than	shift >.
Multiply	shift *8	Less than	shift <,
Divide	?/	Left parenthesis	shift (9
Exponent	shift ^6	Right parenthesis	shift)0

Example 4.1A Building a Formula

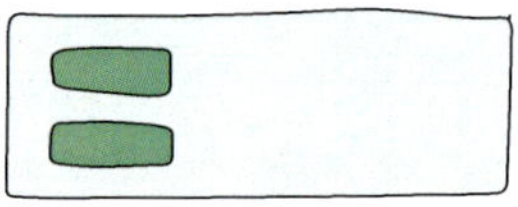

Every formula begins with an equals = sign. In this example, create a formula to add 1 + 2.

Step 1

Open a new spreadsheet and type an equals = sign to start a formula.

	A	B	C	D
1	=			
2				

Step 2

Type the expression you want to calculate.

	A	B	C	D
1	=1+2			
2				

Step 3

Press enter or tab to calculate the expression.

Note: Do not click on another cell to get out of the selection. It will create a cell reference—a topic covered later in the chapter.

	A	B	C	D
1	=1+2			
2				

Result

Your spreadsheet will make the calculation and display the result.

	A	B	C	D
1	3			
2				

The Formula Bar

Above your spreadsheet is the formula bar, which displays the formula in the selected cell. Use the formula bar to review and edit your work.

fx	

	A	B	C	D
1				
2				

How the Formula Bar Works

When a cell is selected, the formula in that cell will appear in the formula bar. The result of the calculation will appear in the cell.

Formula →	fx	=1+2

		A	B	C
Result →	1	3		
	2			

Example 4.1B Using the Formula Bar

Once created, you can edit your formula. In this example, use the formula bar to change the formula from "1+2" to "1+3."

Step 1

Complete the steps in "Example 4.1A Building a Formula" on page 52. Click the cell to edit it.

	A	B	C	D
1	3			
2				

Step 2

The formula in that cell will appear in the formula bar. Click on the formula bar to edit the formula.

fx	=1+2

	A	B	C	D
1	3			
2				

Step 3

Change the formula to "1+3."

fx	=1+3

	A	B	C	D
1	3			
2				

Step 4

Press enter or tab to complete the change and perform the calculation.

Result

The formula will recalculate and display the result in the cell. The formula remains behind the scenes, as seen in the formula bar.

fx	=1+3

	A	B	C	D
1	4			
2				

4.2 Cell References

One major advantage of spreadsheets is the ability to use cell references instead of typing in specific numbers. Each cell has a specific name based on its column letter and row number.

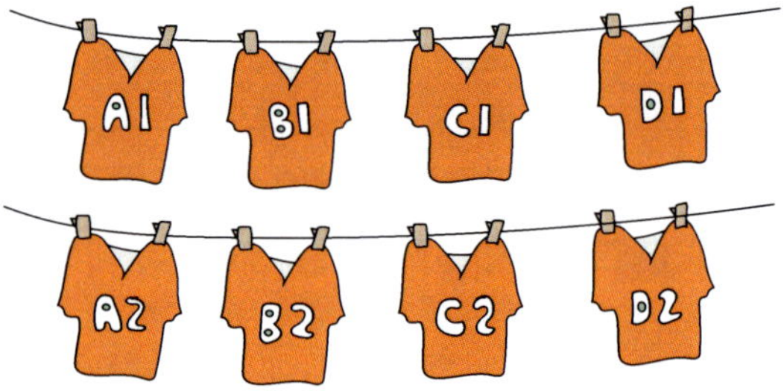

Note: Sports teams wear numbered jerseys to help fans identify each player on the field. Spreadsheets are similar: each cell in a spreadsheet has a specific name so you can stay organized.

Cell Names

Cell names are based on the column letter and the row number. In this case, column B and row 3 are highlighted, so the cell name is B3. *Note: The letter always goes first and the number second.*

	A	B	C	D
1				
2				
3		B3		

Types of References

There are several ways to reference cells in a formula.

Cell Reference

Make a reference to a single cell.

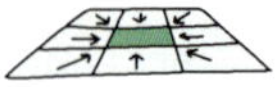

Fixed Reference

Use the same cell reference for several calculations.

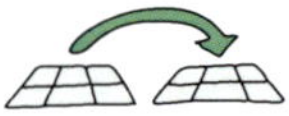

Sheet Reference

Make a reference to a cell in another sheet.

Example 4.2A Creating Cell References

You can create formulas that refer to values in other cells. In this example, add 1+2 using cell references.

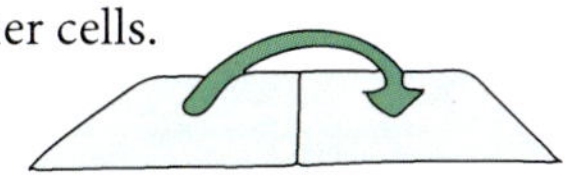

Step 1

Open a new spreadsheet and enter sample values, such as these shown here.

	A	B	C	D
1	1			
2	2			
3				

Step 2

Click on the cell where you want the result of your calculation to go.

	A	B	C	D
1	1			
2	2			
3				

Step 3

Begin the formula by typing an equals = sign.

	A	B	C	D
1	1			
2	2			
3	=			

Step 4

Instead of typing in the number 1, click on it. It will become highlighted and its cell name, A1, will appear in the formula.

	A	B	C	D
1	1			
2	2			
3	=A1			

Step 5

Type a plus + symbol and click on the second number. Its cell name will appear in the formula.

	A	B	C	D
1	1			
2	2			
3	=A1+A2			

Step 6

Press enter or tab to complete the formula and make the calculation. *Note: Do not click on another cell—it will create another cell reference.*

Result

The result of the calculation will be displayed in the cell. The formula will remain behind the scenes, and can be accessed through the formula bar.

fx	=A1+A2

	A	B	C	D
1	1			
2	2			
3	3			

Example 4.2B Editing Referenced Cells

Once you create a formula using cell references, you can change the values in the referenced cells and the result will update automatically. In this example, edit one of the referenced cells from the previous example.

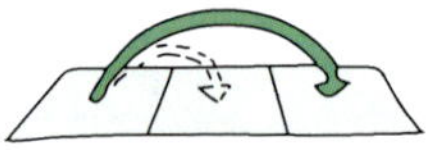

Step 1

Complete the steps in "Example 4.2A Creating Cell References" on page 57. Click on the cell you want to change.

	A	B	C	D
1	1			
2	2			
3	3			

Step 2

Type a new value.

	A	B	C	D
1	7			
2	2			
3	3			

Step 3

Press enter or tab to complete the change.

Result

The result will update, while the formula stays the same.

fx	=A1+A2

	A	B	C	D
1	7			
2	2			
3	9			

The Fill Handle

Builders lay tile by completing a pattern. The spreadsheet fill handle works in a similar way. Using the formulas and values already in place, the fill handle completes the pattern by filling in the rest of the spreadsheet with those formulas or numbering patterns.

Example 4.2C Using the Fill Handle

The fill handle works by copying formulas throughout a spreadsheet. In this example, create a formula that multiplies each number listed in column A by 2.

Step 1

Open a new spreadsheet and enter sample values in column A, such as those shown below.

	A	B	C	D
1	10			
2	20			
3	30			

Step 2

Click on cell B1 and type an equals = sign to start the formula.

	A	B	C	D
1	10	=		
2	20			
3	30			

Step 3

Click on the value in cell A1 to create a cell reference in your formula.

	A	B	C	D
1	10	=A1		
2	20			
3	30			

Step 4

Multiply * it by 2.

	A	B	C	D
1	10	=A1*2		
2	20			
3	30			

Step 5

Press enter or tab to complete the formula.

Step 6

Your spreadsheet will make the calculation and display the result. Click on the answer in cell B1.

fx	=A1*2

	A	B	C	D
1	10	20		
2	20			
3	30			

Step 7

Click and hold the fill handle located at the bottom right corner of the cell and drag down. *Note: If you are using a touchscreen device, refer to "Example 3.3A Creating a List of Numbers" on page 47 for instructions on how to use the fill handle with a touchscreen.*

	A	B	C	D
1	10	20		
2	20			
3	30			

	A	B	C	D
1	10	20		
2	20			
3	30			

Result

The fill handle completes the rest of the calculations by copying the formula, which multiplies the values in column A by 2, and displays the results in the corresponding cells in column B.

	A	B	C	D
1	10	20		
2	20	40		
3	30	60		

How the Fill Handle Works

In the previous example, the fill handle copied the formulas to other cells. Take a closer look at what happens to the formulas when you use the fill handle.

	A	B
1	10	=A1*2
2	20	=A2*2
3	30	=A3*2

When you use the fill handle, the formula stays the same, but the cell reference changes.

Fixed Cell References

Sometimes you may not want the cell references to change when using the fill handle. Instead, you can create a fixed cell reference by placing dollar $ signs before both the column letter and the row number in your cell reference.

```
=$column letter$row number
```

Example 4.2D Creating a Fixed Cell Reference

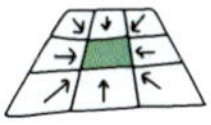

In this example, create a formula that adds the value in A1 to the values in column B, and displays the results in column C.

Step 1

Open a blank spreadsheet and enter values, such as those shown below.

	A	B	C	D
1	5	10		
2		20		
3		30		

Step 2

Select cell C1 and type an equals = sign to start a formula.

	A	B	C	D
1	5	10	=	
2		20		
3		30		

Step 3

Select cell A1.

	A	B	C	D
1	5	10	=A1	
2		20		
3		30		

Step 4

Since you want to add the value in A1 to each of the values in column B, make it a fixed reference. Click on the formula bar to edit the cell reference.

fx	=A1

	A	B	C	D
1	5	10	=A1	
2		20		
3		30		

Step 5

Type dollar $ signs in front of both the column letter and the row number.

fx	=$A$1

	A	B	C	D
1	5	10	=$A$1	
2		20		
3		30		

Step 6

Type a plus + symbol and click on cell B1. Since you want this cell reference to change as you drag down the fill handle, do not use dollar $ signs in it.

fx	=$A$1+B1

	A	B	C	D
1	5	10	=$A$1+B1	
2		20		
3		30		

Step 7

Press enter or tab to complete the formula.

Step 8

Select the result in cell C1.

	A	B	C	D
1	5	10	15	
2		20		
3		30		

Step 9

Click and hold the fill handle located at the bottom right-hand corner of the cell and drag it down. *Note: If you are using a touchscreen device, refer to "Example 3.3A Creating a List of Numbers" on page 47 for instructions on how to use the fill handle with a touchscreen device.*

	A	B	C	D
1	5	10	15	
2		20		
3		30		

	A	B	C	D
1	5	10	15	
2		20		
3		30		

Result

The rest of the calculations are made by copying the formula.

	A	B	C	D
1	5	10	15	
2		20	25	
3		30	35	

How Fixed Cell References Work

In the previous example, you created a fixed cell reference. Compare what happens in a formula with a fixed cell reference and a normal cell reference.

	A	B	C
1	5	10	=$A$1+B1
2		20	=$A$1+B2
3		30	=$A$1+B3

The fixed cell reference will stay the same as the fill handle is dragged down, while the normal cell reference will change.

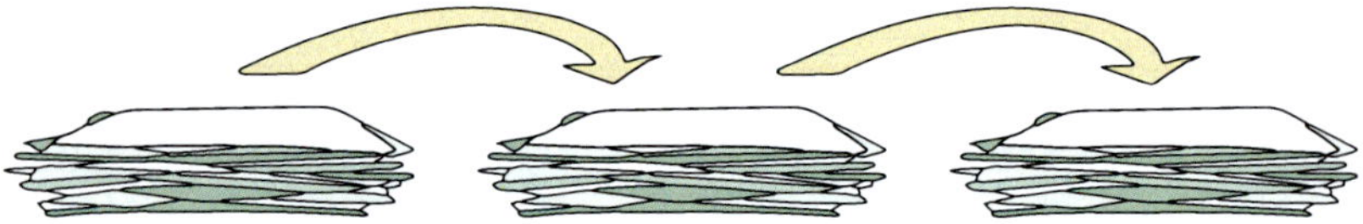

Sheet References

When creating formulas, you can also make references to cells in other sheets. In order to create a formula with a sheet reference, write the name of the sheet, followed by an exclamation point ! and the cell name of the reference you want to make.

=sheet name!cell name

Example 4.2E Creating Sheet References

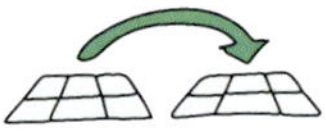

You can create formulas with information from other sheets. In this example, use sheet references to create a formula that multiplies a value in Sheet1 by 2 and displays the result in Sheet2.

Step 1

Open a new spreadsheet and type a value in the first cell.

	A	B	C	D
1	15			
2				

Step 2

Press [enter] or [tab] to complete the entry.

Step 3

Click on the plus [+] symbol next to Sheet1.

	A	B	C	D
1	15			
2				
	Sheet1 +			

Step 4

A new sheet will be created with the default name Sheet2. Click on the first cell.

	A	B	C	D
1				
2				
	Sheet1	Sheet2		

Step 5

Type an equals = sign to begin a formula.

	A	B	C	D
1	=			
2				
	Sheet1	Sheet2		

Step 6

To create a sheet reference, click on Sheet1.

	A	B	C	D
1	=			
2				
	Sheet1	Sheet2		

Step 7

Sheet1 will open and the formula bar will display the name of the sheet followed by an exclamation point !.

fx	=Sheet1!

	A	B	C	D
1	15			
2				
	Sheet1	Sheet2		

Step 8

Click on cell A1 to create the reference. The formula bar will display its cell name.

fx	=Sheet1!A1

	A	B	C	D
1	15			
2				
	Sheet1	Sheet2		

Step 9

Multiply `*` it by 2.

fx	=Sheet1!A1*2			
	A	B	C	D
1	15			
2				
	Sheet1	Sheet2		

Step 10

Press `enter` or `tab` to complete the formula.

Result

The result will be calculated and displayed in Sheet2.

fx	=Sheet1!A1*2			
	A	B	C	D
1	30			
2				
	Sheet1	Sheet2		

4.3 Copy & Paste Options

The default in a spreadsheet is to copy and paste formulas and their results, but there are options. You can copy the result with or without its formula.

Example 4.3A Paste with Formulas

The default option is to paste formulas. In this example, copy and paste the value in a cell and keep its formula.

Step 1

Open a new spreadsheet and type a formula, such as the one shown below.

	A	B	C	D
1	=1+2			
2				

Step 2

Press enter or tab to complete the formula.

Step 3

Select the cell you want to copy and paste.

	A	B	C	D
1	3			
2				

Step 4

Press ctrl and C on the keyboard to copy the cell.

Step 5

The cell border will become dashed indicating it is copied.

	A	B	C	D
1	3			
2				

Step 6

Click on another cell.

	A	B	C	D
1	3			
2				

Step 7

Press ctrl and V on the keyboard to paste it in the new cell.

Step 8

Notice the formula is still in the formula bar. This means the formula has been pasted into the new cell.

fx	=1+2

	A	B	C	D
1	3		3	
2				

Step 9

Press enter or tab to complete the paste.

Result

The formula and its result will be pasted in the new cell.

fx	=1+2

	A	B	C	D
1	3		3	
2				

Example 4.3B Paste Only Values

You can also choose to paste the result without its formula.

Step 1

Open a new spreadsheet and type a formula, such as the one shown below.

	A	B	C	D
1	=1+2			
2				

Step 2

Press enter or tab to complete the formula.

Step 3

Select the cell you want to copy.

	A	B	C	D
1	3			
2				

Step 4

Press ctrl and C on the keyboard to copy the cell.

Step 5

The cell border will become dashed, indicating it is copied.

	A	B	C	D
1	3			
2				

Step 6

Click on another cell.

	A	B	C	D
1	3			
2				

Step 7

Press ctrl and V on the keyboard to paste it in the new cell.

Step 8

To remove the formula and keep only the result, click the clipboard icon in the bottom right corner of the new cell.

fx	=1+2

	A	B	C	D
1	3		3	
2				

Step 9

A menu will appear. Click on the [123] icon to paste the result and remove the formula.

	A	B	C	D
1	3		3	
2				

Result

The formula bar will show the value instead of the formula.

fx	3

	A	B	C	D
1	3		3	
2				

Chapter 4 Summary

Formulas in spreadsheets are powerful tools for making calculations quickly. This chapter covered how to use and edit formulas to make a simple calculation in a single cell and how to use cell references to make many repeated calculations.

Chapter 5: Functions

Spreadsheet applications have many functions to make complex calculations. In this chapter, learn how to use functions to harness the power of spreadsheets.

Chapter 5 Contents

Topics	Examples
5.1 Functions with a Range of Cells	Example 5.1A Using the Sum Function
5.2 Function Arguments	Example 5.2A Using the Payment Function Example 5.2B Using the Future Value Function Example 5.2C Using the Present Value Function
5.3 Logical Functions	Example 5.3A Using the IF Function

Functions

Construction equipment greatly increases productivity on the job. Similarly, functions increase the power of spreadsheet formulas, because they go beyond the basic operations of addition, subtraction, multiplication, and division to make more complex calculations.

Types of Functions

Spreadsheet applications have many different functions. There are three main types of functions, based on the information the functions require.

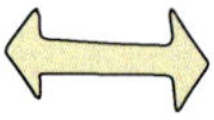

Range

Many functions make calculations based on a range of cells.

Argument

Some functions make calculations based on a list of arguments, or pieces of information.

Logic

Other functions test to see if a cell meets certain criteria.

5.1 Functions with a Range of Cells

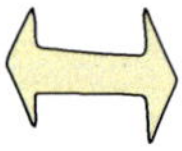

Like formulas, all functions begin with an equals = sign, followed by the name of the function. Then in parentheses ( ), the names of the first cell and the last cell in the range, separated by a colon :.

=function name(first cell:last cell)

List of Functions with a Range of Cells

There are many functions that select a range of cells. Below is a list of some of the most common ones.

Minimum

The `min` function finds the smallest value in a range of data.

Maximum

The `max` function finds the largest value in a range of data.

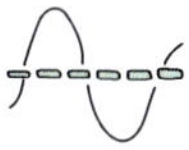

Average

The `average` function calculates the average (mean) of a range of data.

Sum

The `sum` function adds the values in a range of data.

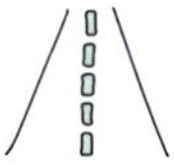

Median

The `median` function finds the middle value in a range of data.

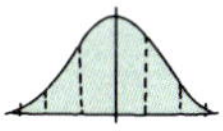

Standard Deviation

The `stdev` function calculates the standard deviation of a range of data.

Example 5.1A Using the Sum Function

In this example, use the `sum` function to add the values in a range of cells.

Step 1

Open a new spreadsheet and enter values, such as those shown below.

	A	B	C	D
1	1			
2	2			
3	3			

Step 2

Select the cell where you want the result of the sum function to go.

	A	B	C	D
1	1			
2	2			
3	3			
4				

Step 3

Begin the function by typing an equals = sign, followed by the word sum. Then open a set of parentheses (.

	A	B	C	D
1	1			
2	2			
3	3			
4	=sum(			

Step 4

Click and hold the first cell in the range and drag down to the last cell. The first and last cell names in your selection will appear in your formula, separated by a colon :. Close the parentheses).

	A	B	C	D
1	1			
2	2			
3	3			
4	=sum(A1:A3)			

Step 5

Press enter or tab on your keyboard to complete the function.

Result

Your spreadsheet will add the values in the range and display the result.

fx	=sum(A1:A3)

	A	B	C	D
1	1			
2	2			
3	3			
4	6			

5.2 Function Arguments

Some functions use arguments instead of a range of cells. An argument is a piece of information needed to make a calculation. For example, the time it takes to repay a loan is an argument. The interest rate would be another argument.

Common Functions with Arguments

Many functions involve arguments. Here is a list of a few common ones.

PMT - Payment

The `pmt` function calculates the payment on a loan.

FV - Future Value

The `fv` function calculates the future value of an investment based on the amount invested today.

PV - Present Value

The `pv` function calculates the value that must be invested today to earn a desired amount in the future.

Creating a Function with Arguments

Start with an equals = sign, followed by the name of the function. Then in parentheses ( ), write the information for each argument separated by commas , .

```
=function name(first argument,second argument,third argument)
```

Common Arguments

Many functions involve arguments. Here is a list of a few common ones.

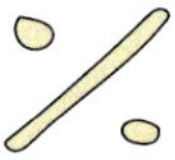

Rate - Rate per Period

The rate is used to calculate interest on an investment or loan.

NPER - Total Number of Periods

The number of compounds in an investment or the total number of payments to repay a loan.

PMT - Payment

Regular contributions to an investment.

PV - Present Value

The current value of an investment or loan.

FV - Future Value

The value of an investment or loan in the future.

Periods per Year

To work with periods in a function, it is important to know the common periods.

Annually

1 time per year

Quarterly

4 times per year

Weekly

52 times per year

Semiannually

2 times per year

Monthly

12 times per year

Daily

365 times per year

Rate per Period

The annual percentage rate (APR) is the interest as a percentage for one year. However, most loans are repaid monthly and investments often compound interest more than once per year.

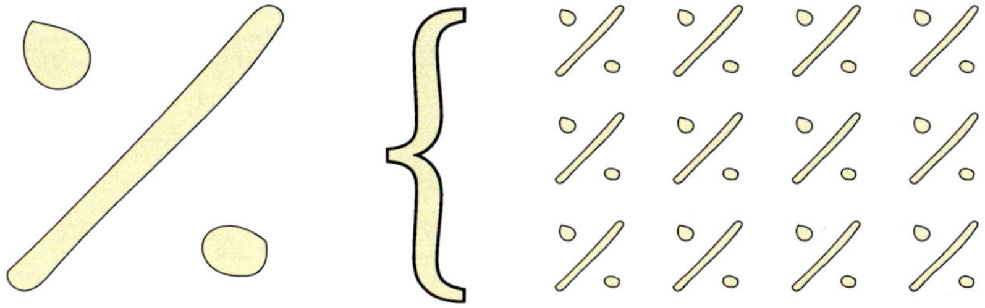

The interest rate for one year

is divided equally among the number of periods per year.

Formatting Money in a Function

When working with the present and future values in a function, there are a few rules to follow for entering dollar amounts into a function.

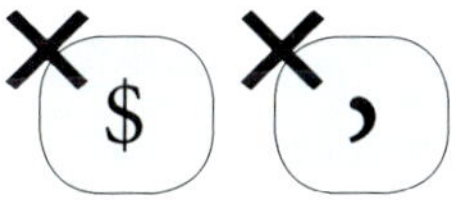

Do not use dollar signs or commas when entering dollar amounts into a function.

Correct = 15000

Incorrect { $15,000 / $15000 / 15,000

The Payment Function

The payment function `pmt` calculates the payment on a loan. In this example, there are three arguments: the rate per period, the total number of periods, and the present value of the loan.

```
=pmt(rate per period,total number of periods,present value of the loan)
```

Example 5.2A Using the Payment Function

Use the payment function [pmt] to calculate the monthly payment on a car loan of $15,000 with an interest rate of 6% to be repaid over 5 years.

Step 1

Open a new spreadsheet and type an equals [=] sign, followed by [pmt], the abbreviation for the payment function. Then open a set of parentheses [(].

	A	B	C
1	=pmt(		
2			

Step 2

The first argument is the rate per period. Type the rate and divide [/] it by 12 months since the car loan is repaid monthly. Then type a comma [,].

	A	B	C
1	=pmt(6%/12,		
2			

Step 3

The second argument is the total number of periods. Type the number of years and multiply [*] it by 12 months per year. Then type a comma [,].

	A	B	C
1	=pmt(6%/12,5*12,		
2			

Step 4

The third argument is the present value of the loan, which is the amount borrowed. Make it negative to indicate debt, and be sure not to include dollar signs or commas in the number. Then close the parentheses [)].

	A	B	C
1	=pmt(6%/12,5*12,-15000)		
2			

Step 5

Press (enter) or (tab) to complete the function and perform the calculation.

Result

The spreadsheet will calculate the monthly payment to repay the $15,000 car loan over 5 years at 6% annual interest.

fx	=pmt(6%/12,5*12,-15000)

	A	B	C
1	$289.99		
2			

The Future Value Function

The future value function calculates the amount an investment will be worth in the future. In this example, there are four arguments: the rate per period, the total number of periods, the payment, and the present value.

=fv(rate per period,total number of periods,payment,present value)

Example 5.2B Using the Future Value Function

Use the future value function to determine the value of an investment if you make an initial investment of $1000 and then invest $5 every day over the course of 40 years with an interest rate of 8% compounded daily.

Step 1

Open a new spreadsheet and type an equals [=] sign followed by [fv], the abbreviation for the future value function. Then open a set of parentheses [(].

	A	B	C
1	=fv(		
2			

Step 2

The first argument is the rate per period. Since interest is compounded daily, divide [/] the rate by 365 days in a year. Then type a comma [,].

	A	B	C
1	=fv(8%/365,		
2			

Step 3

The second argument is the total number of periods. Multiply [*] the number of years by the number of periods per year. Then type a comma [,].

	A	B	C
1	=fv(8%/365,40*365,		
2			

Step 4

The third argument is the payment, which is the amount invested each day. Make it negative [–] because you are investing the money. Then type a comma [,].

	A	B	C
1	=fv(8%/365,40*365,-5		
2			

Step 5

The fourth argument is the present value, which is the initial amount invested. Also make it negative [−] because you are investing the money. Then close the parentheses [)].

	A	B	C
1	=fv(8%/365,40*365,-5,-1000)		
2			

Step 6

Press [enter] or [tab] to complete the function and perform the calculation.

Result

The future value function [fv] calculates the value of the investment after 40 years.

fx	=fv(8%/365,40*365,-5,-1000)

	A	B	C
1	$561,163.58		
2			

The Present Value Function

The present value function calculates the amount of money you need to invest today in order to have a certain amount in the future. In this example, there are four arguments: the rate per period, the total number of periods, the payment, and the future value.

=pv(rate per period,total number of periods,payment,future value)

Example 5.2C Using the Present Value Function

Use the present value function to determine how much to invest today if you want to have $50,000 in 20 years. The interest rate is 7.5% compounded monthly and you plan to only make one initial investment.

Step 1

Open a new spreadsheet and type an equals [=] sign, followed by [pv], the abbreviation for the present value function. Then open a set of parentheses [(].

	A	B	C
1	=pv(		
2			

Step 2

The first argument is the rate per period. Since it is compounded monthly, divide [/] the rate by 12. Then type a comma [,].

	A	B	C
1	=pv(7.5%/12,		
2			

Step 3

The second argument is the total number of periods. Multiply [*] the number of years by the number of periods per year. Then type a comma [,].

	A	B	C
1	=pv(7.5%/12,20*12,		
2			

Step 4

The third argument is the payment. Since you plan to make only one initial investment, there are no additional payments. Type [0] followed by a comma [,].

	A	B	C
1	=pv(7.5%/12,20*12,0,		
2			

Step 5

The fourth argument is the future value, which is the amount you want the investment to be worth in the future. Make it negative [-] since you don't have the money yet. Then close the parentheses [)].

	A	B	C
1	=pv(7.5%/12,20*12,0,-50000)		
2			

Step 6

Press [enter] or [tab] to complete the function to perform the calculation.

Result

The present value function [pv] calculates the amount you need to invest today in order to have $50,000 in 20 years if the investment is compounded monthly at a 7.5% annual interest rate.

fx	=pv(7.5%/12,20*12,0,-50000)

	A	B	C
1	$11,208.71		
2			

5.3 Logical Functions

A logical function tests a cell to determine if it meets certain criteria. The output of the function depends on the results of the test.

Example 5.3A Using the IF Function

The if function checks whether a condition is met, and returns one value if true, and another value if false. For example, tipping is often an if function—if the bill is less than $10, tip $1. If the bill is not less than $10, tip a percentage of the bill.

=if(logical test,value if true,value if false)

Is the bill less than $10?

If yes, tip $1.

If no, tip 20% of the bill.

Step 1

Open a new spreadsheet and enter sample bill amounts, such as those shown below.

	A	B	C	D
1	$12.85			
2	$6.99			
3	$24.31			

Step 2

Select cell B1, type an equals = sign followed by if, and open a set of parentheses (.

	A	B	C	D
1	$12.85	=if(		
2	$6.99			
3	$24.31			

Step 3

The first part of the if function is to create a test. Click on the bill amount in A1 and test to see if it is less than < $10. Then type a comma ,.

	A	B	C	D
1	$12.85	=if(A1<10,		
2	$6.99			
3	$24.31			

Step 4

The second part of the if function is to tell your function what to do if the test is true. If the bill is less than $10, the tip is $1. Type the number 1 followed by a comma ,.

	A	B	C	D
1	$12.85	=if(A1<10,1,		
2	$6.99			
3	$24.31			

Step 5

The third part of the if function is to tell your function what to do if the test is false. In this case, click on the bill in cell A1 and multiply * it by 20% to get the tip. Then close the parentheses).

	A	B	C	D
1	$12.85	=if(A1<10,1,A1*20%)		
2	$6.99			
3	$24.31			

Step 6

Press enter or tab to complete the function.

Step 7

The function will calculate the tip. In this case, the tip is 20% of the bill.

fx	=if(A1<10,1,A1*20%)

	A	B	C	D
1	$12.85	$2.57		
2	$6.99			
3	$24.31			

Step 8

To calculate the rest of the tips, click on the fill handle and drag down.

	A	B	C	D
1	$12.85	$2.57		
2	$6.99			
3	$24.31			

	A	B	C	D
1	$12.85	$2.57		
2	$6.99			
3	$24.31			

Result

The rest of the tip amounts will calculate. Notice how the bill that is less than $10 has a tip of $1. The rest are 20% of the bill.

	A	B	C	D
1	$12.85	$2.57		
2	$6.99	$1.00		
3	$24.31	$4.86		

Chapter 5 Summary

Functions increase the power of spreadsheet formulas by going beyond the basic operations of addition, subtraction, multiplication, and division in order to make more complex calculations. This chapter covered how to use functions to harness the power of spreadsheets.

Chapter 6: Charts

Charts, graphs and diagrams are an excellent way to present spreadsheet data clearly and concisely. In this chapter, learn to create charts and graphs from spreadsheet data.

Chapter 6 Contents

6.1 Prepare Data

Before making a chart in a spreadsheet, set up the data in an organized way. Many charts use two columns of data. The first column is a list of categories and the second is the values associated with those categories.

Example 6.1A Setting Up Data

In this example, organize data in two columns.

Category Heading

Information is often organized into categories. The title of the categories goes in cell A1.

	A	B
1	Category Heading	
2		
3		
4		

Categories

The categories help determine the type of chart to make. Categories can be names, years, numbers, etc. List the categories in column A.

	A	B
1	Category Heading	
2	Category 1	
3	Category 2	
4	Category 3	

Chart Title

The name you want to appear as your chart title goes in cell B1. Descriptive titles work best.

	A	B
1	Category Heading	Chart Title
2	Category 1	
3	Category 2	
4	Category 3	

Values

Each category in column A has a numerical value associated with it. List those values in column B.

	A	B
1	Category Heading	Chart Title
2	Category 1	Value 1
3	Category 2	Value 2
4	Category 3	Value 3

6.2 Create Charts

Once the spreadsheet is created and organized, you can choose from a variety of charts. In this chapter, learn the four most common types of charts.

Common Types of Charts

There are many kinds of charts available in spreadsheet applications. The following is a list of three common charts and graphs.

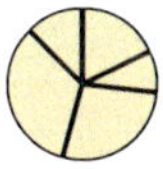

Pie

The pie chart is used to show how parts are related to a total, and often uses percentages.

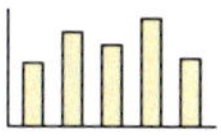

Column

The column chart is useful for comparing different categories.

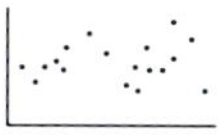

Scatter

The scatter diagram plots points to show the relationship between two variables to see if there is a trend.

Example 6.2A Inserting a Chart

In this example, explore the chart features in Microsoft Excel® or Google Sheets®.

Using Microsoft Excel®

Step 1

Open a new spreadsheet and click on the Insert tab near the top of the window.

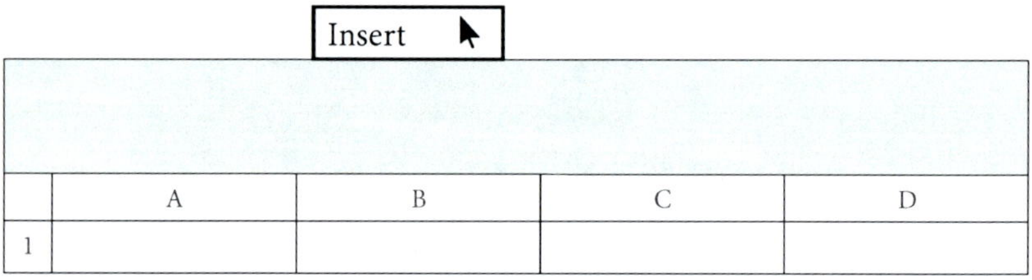

Step 2

Locate the desired chart icon.

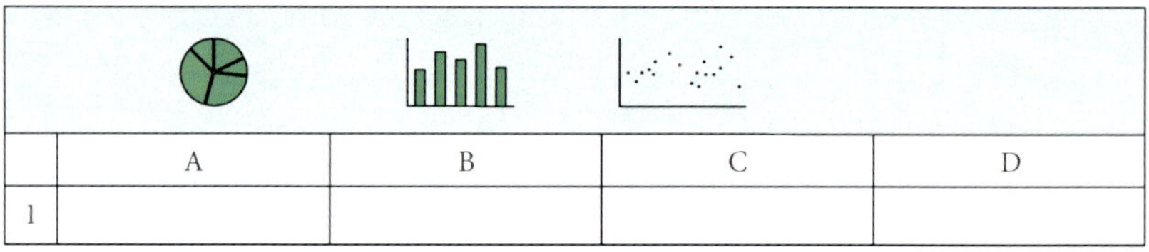

Using Google Sheets®

Step 1

Open a new spreadsheet and click on the Insert tab at the top of the window.

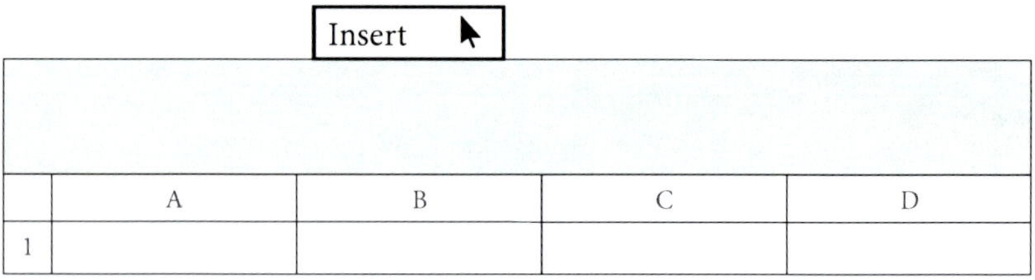

Step 2

Click on (Chart).

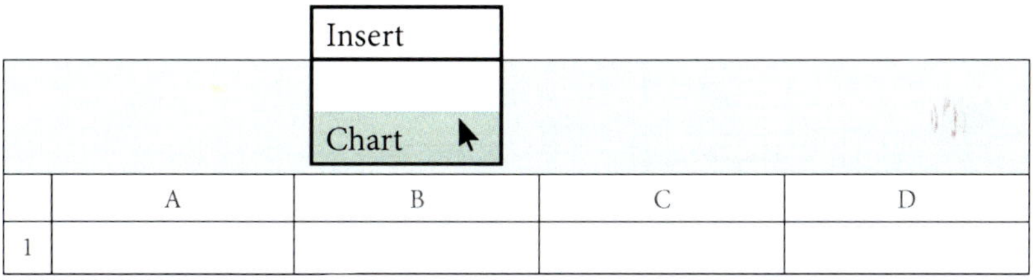

Step 3

A pop-up menu will appear. Choose the desired chart and follow the instructions.

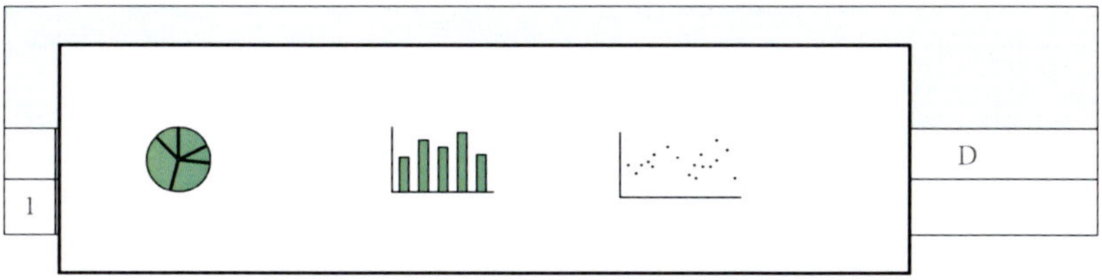

Example 6.2B Creating a Pie Chart

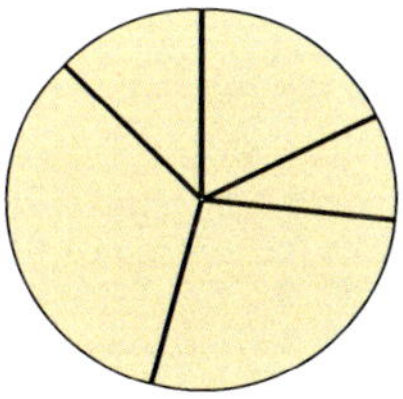

A pie chart is useful to see how the slices of the pie compare to one another and to the whole. In the following example, use a pie chart to compare several expenses within a budget.

Step 1

Open a new spreadsheet and type the data shown below.

	A	B	C
1	Category	Expenses in a Budget	
2	Rent	$950.00	
3	Groceries	$420.00	
4	Utilities	$55.00	
5	Insurance	$110.00	
6	Entertainment	$200.00	

Step 2

Select the data.

	A	B	C
1	Category	Expenses in a Budget	
2	Rent	$950.00	
3	Groceries	$420.00	
4	Utilities	$55.00	
5	Insurance	$110.00	
6	Entertainment	$200.00	

Step 3

Insert a pie chart. *Refer to "Example 6.2A Inserting a Chart" on page 94 for instructions.*

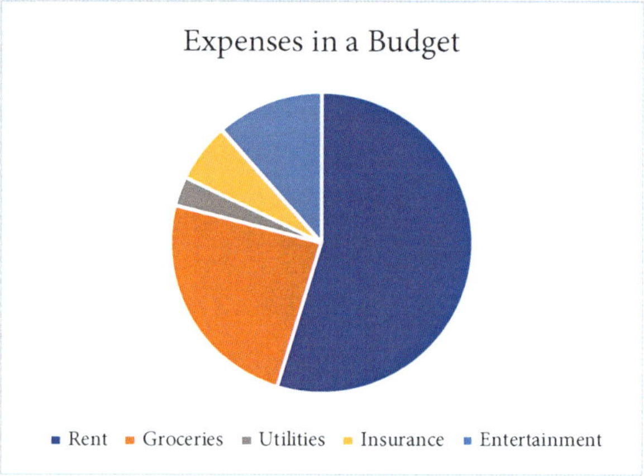

Example 6.2C Creating a Column Chart

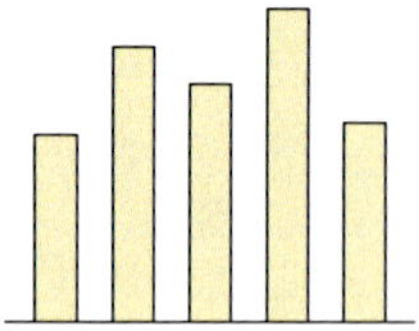

A column chart is useful to compare different categories. In this example, use a column chart to compare the number of employees each business owner currently has.

Step 1

Open a new spreadsheet and type the data shown below.

	A	B	C
1	Business Owner	Employees per Owner	
2	Carlos	450	
3	Susan	220	
4	John	690	
5	Angela	840	
6	Sam	310	

Step 2

Select the data.

	A	B	C
1	Business Owner	Employees per Owner	
2	Carlos	450	
3	Susan	220	
4	John	690	
5	Angela	840	
6	Sam	310	

Step 3

Insert a column chart. *Refer to "Example 6.2A Inserting a Chart" on page 94 for instructions.*

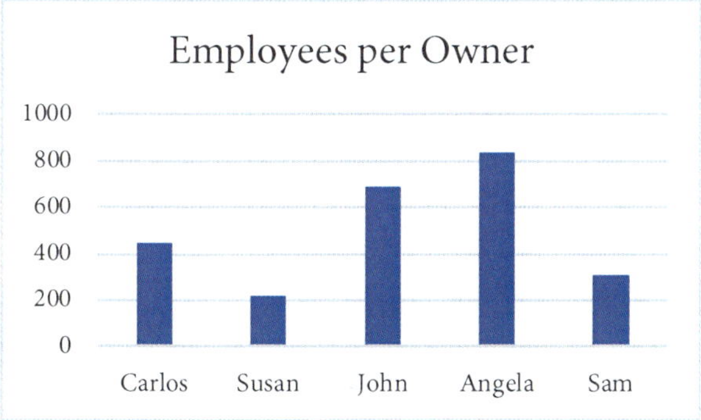

Example 6.2D Creating a Scatter Plot

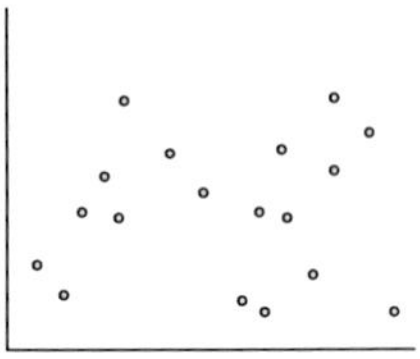

A scatter plot is useful to spot trends and see if there is a correlation between two variables. In this example, use a scatter plot to see if there is a correlation between the hours students spend studying each week and their GPA.

Step 1

Open a new spreadsheet and type the data shown below.

	A	B	C
1	Hours Studied	GPA Based on Hours Studied	
2	0	1.9	
3	5	1.8	
4	10	3.5	
5	15	3.4	
6	20	3.9	

Step 2

Select the data.

	A	B	C
1	Hours Studied	GPA Based on Hours Studied	
2	0	1.9	
3	5	1.8	
4	10	3.5	
5	15	3.4	
6	20	3.9	

Step 3

Insert a scatter plot. *Refer to "Example 6.2A Inserting a Chart" on page 94 for instructions.*

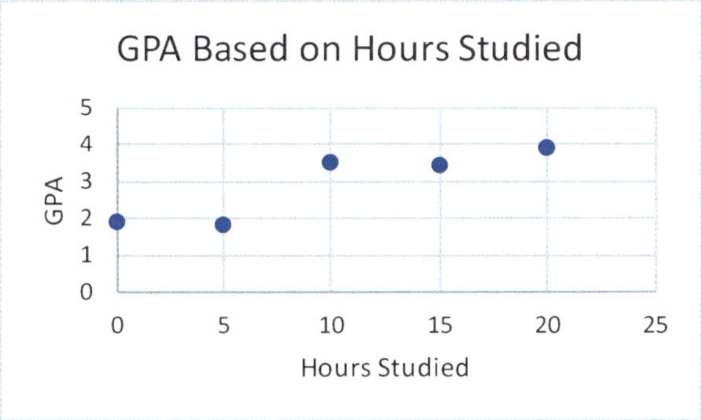

Chapter 6 Summary

Charts present information visually in order to clearly and concisely communicate its meaning or analysis, especially for those who are unfamiliar with the raw data. This chapter covered how to create charts and graphs from spreadsheet data.

About the Author

Award-winning author Pi Luna integrates multiple learning styles to merge logic with meaning. As an artist, designer, and math instructor, Pi believes the best learning happens when both sides of the brain are engaged. She is a nationally certified tutor (CRLA III) and has over 12 years experience tutoring and teaching math at the high school and college level. Pi has a master's degree in Interdisciplinary Art from Goddard College, where she studied the connection between math, art, and psychology. To learn more, visit www.piluna.net.

Made in United States
North Haven, CT
04 November 2026

21726617R00058